3rd ISC International Conference on Intelligent Systems and Control in Fashion and Textile Engineering (ISC-FTE 2022)

MATEC Web of Conferences Volume 359 (2022)

Macau, China
26-28 April 2022

Editor:

Wei D. Deng

ISBN: 978-1-7138-5362-6

Printed from e-media with permission by:

Curran Associates, Inc.
57 Morehouse Lane
Red Hook, NY 12571

Some format issues inherent in the e-media version may also appear in this print version.

For additional information, please contact EDP Sciences – Web of Conferences
at the address below.

EDP Sciences – Web of Conferences
17, Avenue du Hoggar
Parc d'Activité de Courtabœuf
BP 112
F-91944 Les Ulis Cedex A
France

Phone: +33 (0) 1 69 18 75 75
Fax: +33 (0) 1 69 28 84 91

contact-edps@webofconferences.org

Additional copies of this publication are available from:

Curran Associates, Inc.
57 Morehouse Lane
Red Hook, NY 12571 USA
Phone: 845-758-0400
Fax: 845-758-2633
Email: curran@proceedings.com
Web: www.proceedings.com

TABLE OF CONTENTS

The Preface

We are pleased to announce that 2022 3rd ISC International Conference on Intelligent Systems and Control in Fashion and Textile Engineering (ISC-FTE 2022) has been held on April 26-28, 2022, Macau, Special Administrative Region (SAR) of the People's Republic of China.

Macau is a Special Administrative Region (SAR) of the People's Republic of China. Located across the Pearl River estuary from Hong Kong, Macau was until 1999 an overseas territory of Portugal. One of the world's most densely populated spots, Macau generates more revenue from gambling than anywhere else on the planet, including more than seven times the revenue generated by "The Strip" in Las Vegas.

2018 2nd AASRI International Conference on Intelligent Systems and Control (ISC 2018) had been held in Lima, Peru, December 27-29, 2018. ISC 2018 is sponsored by American Applied Sciences Research Institute, USA. ISC 2018 is also technically co-sponsored by University of Wisconsin, USA, University of Gothenburg, Sweden and Aalto University, Finland. ISC 2018 Papers had been published by MATEC Web of Conferences (ISSN: 2261-236X) ,Volume 267, 2019, which had been archived in EDP Sciences Digital Library, been indexed by Ei Compendex and Scopus, and Conference Proceedings Citation Index (CPCI) successfully.

The goal of ISC-FTE 2022 is to bring together the researchers from academia and industry as well as practitioners to share ideas, problems and solutions relating to the multifaceted aspects of Control and Automation, Communications and Information technology, Computer, Sustainable Fashion and Textile Engineering.

Control engineering or control systems engineering or Automation engineering is an engineering discipline that deals with control systems, applying control theory to design equipment and systems with desired behaviors in control environments. The discipline of controls overlaps and is usually taught along with electrical engineering and mechanical engineering at many institutions around the world.

The practice uses sensors and detectors to measure the output performance of the process being controlled; these measurements are used to provide corrective feedback helping to achieve the desired performance. Systems designed to perform without requiring human input are called automatic control systems (such as cruise control for regulating the speed of a car). Multi-disciplinary in nature, control systems engineering activities focus on implementation of control systems mainly derived by mathematical modeling of a diverse range of systems.

Information and communications technology (ICT) is an extensional term for information technology (IT) that stresses the role of unified communications and the integration of telecommunications (telephone lines and wireless signals) and computers, as well as necessary enterprise software, middleware, storage and audiovisual, that enable users to access, store, transmit, understand and manipulate information.

ICT is also used to refer to the convergence of audiovisual and telephone networks with computer networks through a single cabling or link system. There are large economic incentives to merge the telephone network with the computer network system using a single unified system of cabling, signal distribution, and management.

Textile manufacturing is a major industry. It is largely based on the conversion of fibre into yarn, then yarn into fabric. These are then dyed or printed, fabricated into cloth which is then converted into useful goods such as clothing, household items, upholstery and various industrial products. Overall, many things can be made with cotton, not just clothing. Automation, includes the use of various equipment and control systems such as machinery, processes in factories, boilers, and heat-treating ovens, switching on telephone networks, steering, and stabilization of ships, aircraft, and other applications and vehicles with reduced human intervention.

ISC-FTE 2022 is co-sponsored by American Applied Sciences Research Institute, USA, Universidad de Sevilla, North America of Alumni Association, USA, ISC-FTE 2022 is also Technically Co-sponsored by Yangon ACM Chapter, Myanmar.

ISC-FTE 2022 received 141 submissions. All full submissions were peer reviewed and evaluated based on originality, technical and research content depth, correctness, relevance to conference, contributions, and readability. Submissions were chosen based on technical merit, interest, applicability, and how well they fitted a coherent and balanced technical program. Finally, only 35 best papers were selected as regular papers with the acceptance rate at about 24.8%. iThenticate software was also used for plagiarism detection to ensure the originality of ISC-FTE 2022 proceedings.

The success of this truly international conference is attributed to the efforts of the organizing committee. We also want to thank all the invited speakers, oral and poster presenters, reviewers of manuscripts, and participants of the conference who have contributed to the success of this international conference.

We would like to express our deep appreciation and gratitude to all the people who review the submissions. Then we would also show our sincere thanks to all staff without whose help we can't carry out the important steps during the process of conference preparation. In addition, thanks for your support and participation.

Sincerely

Publication Co- chairs
Dr. Wei David Deng, American Applied Sciences Research Institute, USA
Dr. Biao Harry Zhang, Universidad de Sevilla, North America of Alumni Association, USA
Dr. Riza Esa ,Yangon ACM Chapter, Myanmar

edp sciences

MATEC Web of Conferences

Statement of Peer review

In submitting conference proceedings to MATEC Web of Conferences, I certify to the Publisher that I adhere to the **Policy on Publishing Integrity** of the journal in order to safeguard good scientific practice in publishing.

1. All articles have been subjected to peer review administered by the proceedings editors.
2. Reviews have been conducted by expert referees, who have been requested to provide unbiased and constructive comments aimed, whenever possible, at improving the work.
3. Proceedings editors have taken all reasonable steps to ensure the quality of the materials they publish and their decision to accept or reject a paper for publication has been based only on the merits of the work and the relevance to the journal.

Title, date and place of the conference

2022 3rd ISC International Conference on Intelligent Systems and Control in Fashion and Textile Engineering (ISC-FTE 2022) ,April 26-28, 2022, Macau, Special Administrative Region (SAR) of the People's Republic of China

Proceedings editor(s):

Wei David Deng

Date and editor's signature

MATEC Web of Conferences **359**, 01001 (2022) https://doi.org/10.1051/matecconf/202235901001
ISC-FTE 2022

Emotional Labor and Education Career Development in the Age of Big Data

Fei Qi, *Lei* Zhang

Chengde Medical University, Hebei, China, 067000

Abstract. With the development of psychology research field, researchers and educators increasingly pay attention to teachers' emotional labor in the workplace. This paper mainly studies the development of emotional labor language vocational education for university teachers in the workplace in the era of big data. Through the class questionnaire survey, it is found that the emotional labor level of college teachers is significantly higher than the norm of emotional labor of primary and secondary school teachers in China. Positive and negative emotional experience exist simultaneously. Teachers' professional self-identification level and personal sense of achievement are good, and job burnout of different degrees is common. In the process of emotional labor, the deep acting strategy was mainly adopted and the level of emotional labor perception was better. The emotional labor of teachers in colleges and universities can be jointly assisted by enhancing their emotional restraint, clarifying the rules of emotional expression and improving the emotional labor environment.
Keywords: Big Data, College Teaching, Emotional Labor, Education Career Development.

1 Introduction

Teacher's emotional labor refers to the emotional state that teachers work in accordance with laws and regulations and relevant requirements in teaching activities. At present, researchers have increasingly paid attention to the importance of teachers' emotional labor. Studies have shown that negative emotional labor will cause job burnout and reduce teachers' job satisfaction [1]. The pressure that teachers face mainly includes work pressure, social pressure and family pressure. In the teaching work, teachers face great pressure for a long time, mainly dealing with the inspection of superior departments, dealing with the relationship between colleagues, student performance, dealing with student problems, dealing with the relationship with parents, dealing with the relationship with leaders, curriculum reform requires teachers to constantly improve the knowledge and skills of the pressure, title pressure, etc. Social pressure mainly includes mortgage pressure, social pressure on teachers' role expectation, salary pressure, etc. Family pressure includes marriage pressure, children's education pressure, parental support pressure and so on; Negative work pressure will bring serious negative effects to teachers. If they cannot effectively deal with the negative pressure in work, the exhaustion of emotion, attitude and behavior caused by the long-term experience of pressure is also called job burnout [2-3]. With the new curriculum reform, the requirements of social development on teachers are gradually increasing. In order to better adapt to the requirements of educational development, the working pressure of teachers is also increasing, which is an urgent aspect that needs to be studied.

Scholars have conducted in-depth research on teachers' emotional labor, and there are mainly two views at present. First, teachers' emotional labor is a process or behavior of internal psychological adjustment. This view holds that emotional labor is a kind of emotional behavior that conforms to specific situational rules and is made by individuals in order to get paid. It is a profession that requires constant emotional labor, which is very consistent with the characteristics of emotional labor [4]. Another view is that teachers' emotional labor is the external state of teachers. The second view defines emotional labor as "an individual achieves the desired emotional performance of the organization through their own emotional adjustment, including external performance and internal different feelings" [5]. Scholars have gradually enriched the research on the connotation of teachers' emotional labor and defined the concept of teachers' emotional labor from different perspectives. However, at present, there is no unified definition of teachers' emotional labor [6].

The research on teachers' emotional labor is helpful to guide teachers to effectively regulate their emotions, alleviate the negative effects of negative emotions, and thus contribute to the improvement of teachers' mental health.

2 Emotional labor of college teachers in the age of big data

2.1 Education in the big data Era

Under the guidance of the national big data technology construction, China's higher vocational education is also undergoing a series of changes imperceptibly. With the construction of digital campus, safe campus, mobile APP, MOOC education, "flipped classroom" Weibo, WeChat and other derivatives, students are provided with a broader space for network communication. The university education environment is also trying to change in a digital way. The coming of the era of big data has provided new technical support and opened up new functions for higher education. At the same time, it also poses a challenge to the work of higher education [7-8].

First, it is beneficial to realize the integration of resources in higher education. In the era of big data, educators, libraries, funding departments, educational administration systems, logistics departments and other departments related to student education can use the big data platform to achieve information sharing. Through the collection, analysis and integration of related data scattered among various departments, such as students' consumption on campus, books borrowed, grades, scholarships, etc., big data excavates the situation and behavior dynamics of individual and group students, which is convenient for unified deployment and coordinated promotion of student education.

Second, it is beneficial to broaden the new channels of higher education. Big data provides technological advantages for ideological and political education. In the process of educational work, in addition to the traditional way of work, we can use the Internet, mobile APP, MOOC education, "flipped classroom" microblog, WeChat and other means popular with vocational students to provide students with network communication space, and enhance the attraction and appeal of educational work. At the same time, educators use the big data platform to integrate and analyze the data of students in the network space to help them have a comprehensive and objective understanding of students' thoughts and behaviors, so as to realize that the education work is close to students, close to reality and close to life.

Third, it is beneficial to enhance the effectiveness of college education. In college education, big data can be used to collect students' campus behavior data in a timely and comprehensive way, and to process and mine the scattered mass data, so as to realize the tracking and prediction of students' behavior dynamics. At the same time, big data also provides the possibility to carry out personalized education, thus enhancing the effectiveness of education [9].

2.2 College labor in college teachers

(1) Definition of emotional labor

With the in-depth development of research in the field of psychology, more and more scholars pay attention to the role of emotion in individual life and work. Any kind of work has specific requirements for emotional expression, so individuals will have emotional labor in work [10]. Emotional labor, or emotional work, is an emotional labor. Emotional labor is defined by American scholars as the management of emotional expression by employees in various ways, which can be observed through gestures or facial expressions. Viewing the interaction between employees and customers from the perspective of dramatic theory, the customer is the audience, the employee is the actor, and the workplace is the stage. She believes that the management of employees' emotions plays a key role in their development in the workplace. The process of emotional labor includes three necessary conditions: emotional labor occurs in individual voice-to-voice or face-to-face communication; in the process of direct communication, individuals produce emotional labor with the purpose of triggering a certain emotional state of others through their emotional expression. The organization where they work has certain requirements on the emotions of employees [11-12].

(2) Structure of emotional labor

Some scholars put forward the two-dimensional theory of emotional labor according to the individual's different focus of attention in the workplace. They divide emotional labor into "work-oriented emotional labor" and "employee-oriented emotional labor". Work-oriented emotional labor mainly describes the requirements of the organization on the emotional performance of employees, which focuses on the emotional requirements of the organization and the work on the service and communication process. Employee-oriented emotional labor describes the psychological and behavioral processes corresponding to the performance of the emotional requirements of the organization and work requirements. It focuses on the process of individual emotional disorders and emotional adjustment.

(3) Influencing factors and effects of teachers' emotional labor

Summarizing the results of previous studies on influencing factors of teachers' emotional labor, it mainly includes the following aspects: individual, organizational and environmental factors.

From the perspective of individual factors, teachers' gender, educational background, teaching age, professional title and marital status are important factors affecting teachers' emotional labor.

In addition to subjective factors, emotional labor is also affected by objective factors. People's emotional labor will have different feelings depending on the objective environment. From the perspective of environmental factors, leadership style, peer relationship, teacher-student relationship and so on all affect teachers' emotional labor, can affect teachers' attitude and behavior, and have an impact on their job satisfaction.

From the perspective of organization, the influence of emotional labor on teachers includes the influence on teachers' organization and emotional transmission. Deep behavior will have a positive impact on teachers' emotional transmission, while surface behavior will have a negative impact. Those who are less motivated to work and cannot fully agree with organizational rules are more likely to use surface acting.

From the perspective of teachers, emotional labor affect teacher basically has the following several aspects: teachers will have no discipline students in teaching activities, teachers will suppress anger to show patience and politeness, or when its not agree with the way the school of management, and can't change, and in constant "depressed mood" behavior, the passage of time will produce job burnout. The results show that if teachers adopt more deep behaviors to manage their emotions in their daily work, their enthusiasm will be improved.

3 Survey on emotional labor of college teachers

3.1 Questionnaire survey

In order to explore the relationship between the emotional labor of college teachers and the workplace, and to acquire first-hand data and information, teachers from 3 colleges and universities in the city were randomly selected as subjects. The paper questionnaire was issued to college teachers on emotional labor, and SPSS20.0 software was used to conduct statistical analysis on the collected data.

3.2 Emotional labor scale

The original scale of the questionnaire used in this study to measure the emotional labor of college teachers consists of 27 items. In this study, one item whose factor load is less than 0.5 is deleted. The scale includes four dimensions: emotional expression (including 6 items), surface acting (including 7 items), emotional deviation (including 6 items) and deep acting (including 7 items). Autonomous emotional expression, in which an individual naturally displays the emotions he or she experiences, does not require extra effort; surface acting, the individual through the disguise of their inner true feelings and express the emotions stipulated by the organization; emotion deviation, the individual truly shows the emotions experienced internally, but the emotions displayed are contrary to the organizational requirements. Deep acting, through the adjustment of their inner feelings, the individual shows the appropriate emotions required by the organization.

3.3 Mathematical statistics

In this paper, SPSS20.0 data statistics software was used to code and input the valid data obtained from the questionnaire, and further reliability and validity test, descriptive statistics, t-test, correlation analysis and regression analysis were carried out. The t-test formula used in this paper is as follows:

$$t = \frac{\overline{X} - \mu}{\frac{\sigma x}{\sqrt{n-1}}} \tag{1}$$

$$t = \frac{\overline{x_1} - \overline{x_2}}{\sqrt{\frac{\sigma_{x_1}^2 + \sigma_{x_2}^2 - 2\gamma\sigma_{X1}\sigma_{X2}}{n-1}}} \tag{2}$$

4 Questionnaire of emotional labor questionnaire

4.1 General overview of emotional labor of college teachers

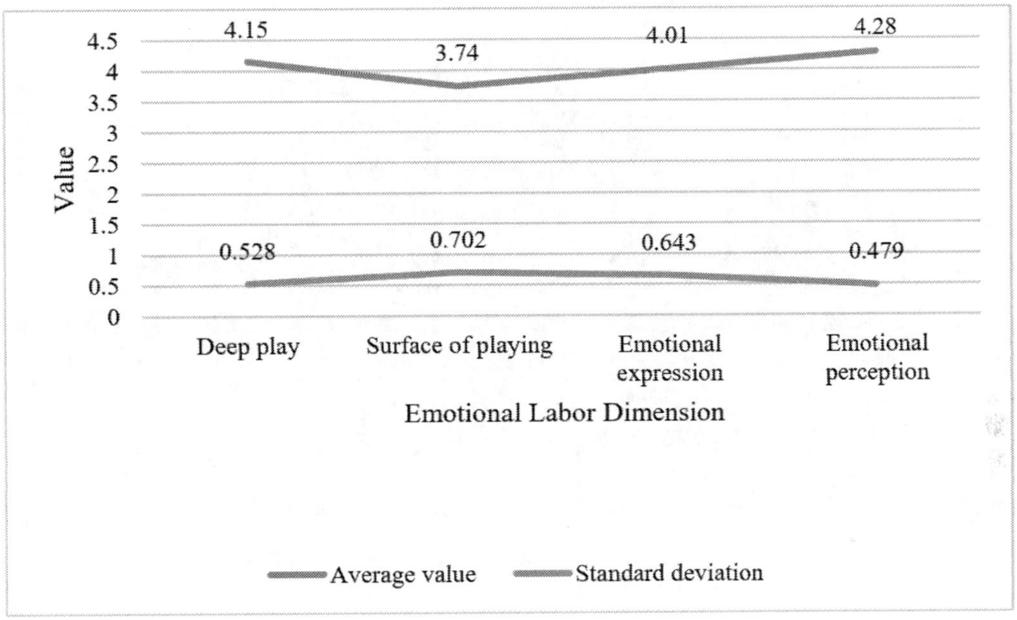

Fig.1. The general situation of emotional labor of college teachers.

As shown in Figure 1, the average score of emotional labor of college teachers is 4.15 for deep acting, 3.74 for surface acting, 4.01 for emotional expression, and 4.28 for emotional deviation. Is on the high side, which proves that college teachers are engaged in high intensity emotional labor in their teaching work.

4.2 Statistics on emotional labor description of college teachers

Table 1. Statistical table of emotional labor description of college teachers.

	Min value	Max value	Average value	Standard deviation
Deep play	3	6	4.15	0.479
Surface of playing	2	6	3.74	0.702
Emotional expression	3	6	4.01	0.643
Emotional perception	3	6	4.28	0.528

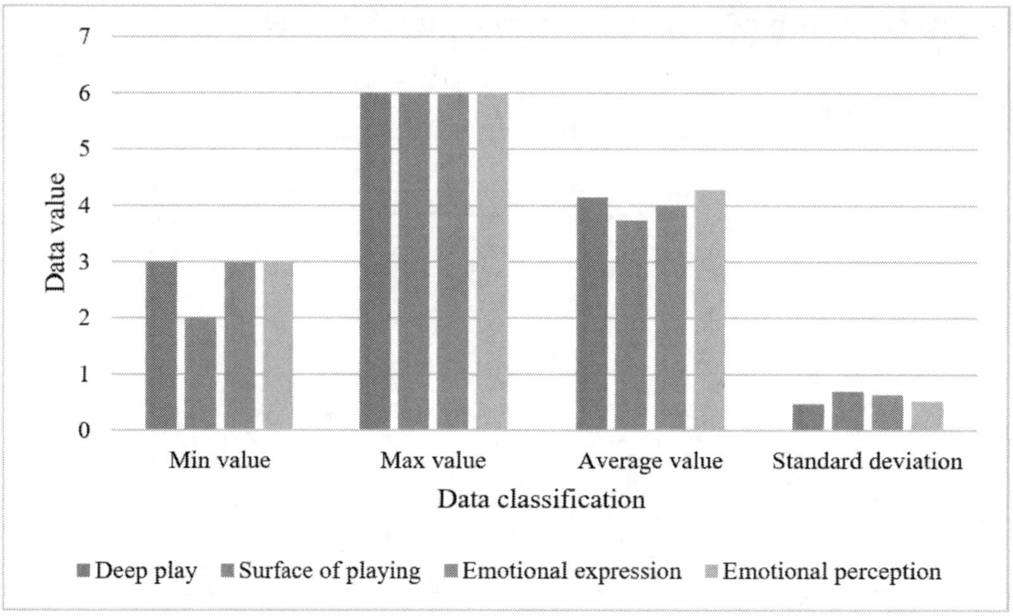

Fig.2. Statistical figure of emotional labor description of college teachers.

As shown in Table 1 and Figure 2, the highest average value is "perceived emotional labor". College teachers generally believe that their emotional labor plays an important role in building harmonious teacher-student relationship, establishing a good teacher image, showing teachers' caring behavior, handling interpersonal relationship and teaching work. In terms of emotional labor strategies, college teachers mostly use deep acting strategies. When personal real emotional experience is inconsistent with the emotions needed to be expressed in teaching work, teachers will adjust and change emotional cognition and external emotional expression. In addition to the deep acting strategy, teachers would also express their real emotional feelings in front of the interactive objects, while the surface acting strategy was the lowest in average value, and college teachers would also use "duplicity" to cover up their real emotional feelings. In general, the average value of the choice and use of various strategies of university teachers is at a high level, belonging to the group of high emotional labor.

4.3 Suggestions on educational career development

(1) Clarifying the emotional expression rules of college teachers

Emotional expression rules are the core elements of emotional labor and the principles and basis for teachers to choose emotional labor strategies. Correct understanding of emotional expression rules is helpful for teachers to express appropriate emotions in their work, shape their self-image, effectively promote the development of teaching work, facilitate interpersonal communication in work, and help the school to achieve the goal of educating students. The rules of emotional expression in most schools are vague and general, which are hidden in tangible or intangible contents such as school culture, school rules and motto, or similar "norms of teachers' conduct" in a way of "unspoken rules". Teachers' emotional expression rules are ubiquitous, but they have not been formally "expressed". Once these rules are internalized into a part of the teacher's personal belief, the teacher will automatically adjust his teaching behavior according to the requirements of the emotional expression rules. Therefore, the construction of emotional expression rules has great significance and influence on teachers' emotional labor, on the smooth development of teaching work and on the realization of teaching objectives.

(2) Improve the emotional labor environment of college teachers

Through the comprehensive analysis of the research results, it can be seen that at present, teachers in China are engaged in high intensity emotional labor, a large number of teachers have negative emotions in daily teaching work, and the problem of job burnout is common. Through in-depth analysis of the interview results, it can be found that some teachers say that when they see students' progress, get recognition from leaders, support from parents and understanding from colleagues, they will feel much better and their emotional problems will be alleviated to a certain extent. From the perspective of resource conservation theory, to some extent, these external resources supplement the emotional resource consumption caused by emotional labor. Aiming at the resource and cultural dilemmas faced by teachers in their actual work, improving the emotional labor environment of college teachers is of great significance and influence to the realization of teaching objectives.

5 Conclusions

Individuals experience a variety of complex emotions in the work of working closely with people. Teachers serve students, and each teacher faces at least dozens of students every day. As teachers are in close contact with students, there is no doubt that teachers will also have complex emotional experience. Reasonable adjustment of their emotions, through the way of reevaluation to understand students' behavior, can effectively adjust the negative emotions. Teachers need to adopt multiple and effective ways to regulate negative emotions at work. There is no best or worst emotion adjustment strategy. Effective coping strategies depend on the ability of individuals to adjust themselves in different situations. The best way to regulate emotions is to use a combination of effective coping styles.

References

1. Joanne, Cliffe. Emotional intelligence testing for headteachers: globalization or lost in translation[J]. Journal of Professional Capital and Community, 2018, 3(3):223-240.
2. Larson, Eric B. Clinical Empathy as Emotional Labor in the Patient-Physician Relationship[J]. Jama the Journal of the American Medical Association, 2016, 293(9):1100-6.
3. Ac Heson K, Taylor J, Luna K. The Burnout Spiral: The Emotion Labor of Five Rural U.S. Foreign Language Teachers[J]. Modern Language Journal, 2016, 100(2):522 - 537.
4. Mauno S, Ruokolainen M, Kinnunen U, et al. Emotional labour and work engagement among nurses: examining perceived compassion, leadership and work ethic as stress buffers[J]. Journal of Advanced Nursing, 2016, 72(5):1169-1181.
5. Shim J M. Teacher as a Foreigner in Critical Multicultural Teacher Education[J]. Nephron Clinical Practice, 2019, 14(2):117-136.
6. Yin H. Knife-like mouth and tofu-like heart: emotion regulation by Chinese teachers in classroom teaching[J]. Social Psychology of Education, 2016, 19(1):1-22.
7. Eichhorn S, Matkin G W. Massive Open Online Courses, Big Data, and Education Research[J]. New Directions for Institutional Research, 2016, 2015(167):27-40.
8. Pence H E, Antony J. Williams. Big Data and Chemical Education[J]. Journal of

Chemical Education, 2016, 93(3):504-508.

9. Horn V, Darrell J. Opinion: Big data biomedicine offers big higher education opportunities[J]. Proceedings of the National Academy of Sciences, 2016, 113(23):6322-6324.
10. Zikhali J, Perumal J. Leading in Disadvantaged Zimbabwean School Contexts: Female School Heads' Experiences of Emotional Labour[J]. Educational Management Administration & Leadership, 2016, 44(3):347-362.
11. Ye H L, Chelladurai P. Affectivity, Emotional Labor, Emotional Exhaustion, and Emotional Intelligence in Coaching[J]. Journal of Applied Sport Psychology, 2016(2):00-00.
12. Yoo J. Perceived customer participation and work engagement: the path through emotional labor[J]. International Journal of Bank Marketing, 2016, 34(7):1009-1024.

MATEC Web of Conferences **359**, 01002 (2022) https://doi.org/10.1051/matecconf/202235901002
ISC-FTE 2022

Research on Multi-modal Teaching of Graduate Students Academic English Writing Based on Information Technology

Meiyi Chen

Hubei University of Arts and Science, Hubei, Xiangyang, 441000, China

Abstract. In the process of innovating the English teaching system, multi-modal teaching is very representative and promotes the digital and three-dimensional development of English teaching. The teaching of academic English writing is of great practical significance for graduate students to carry out academic research and express their academic views independently. In the traditional teaching of graduate students' Academic English writing, the content of the teaching is relatively single, and the teaching process lacks interaction, which affects the continuous improvement of graduate students' Academic English writing. Taking the application of information technology as the research background, this paper puts forward the teaching strategy of multi-modal academic English writing for graduate students.

Keywords: Information technology, multi-modality, graduate students, academic English, writing teaching.

With the development of academic English and educational technology, the teaching of academic English writing has always been a difficult point. At the same time, with the in-depth study of Multimodal Discourse Analysis, multimodal teaching model has been applied to college English teaching. By combining the multi-modal teaching model with the teaching of academic English writing, and using the multi-media resources and the group cooperation teaching method, the symbolic mode and the sensory mode are embodied in the teaching process of academic writing, it arouses students' interest in academic English writing and improves teachers' teaching quality.

1 Analysis of multi-modal teaching mode

Multi-modal teaching belongs to the emerging teaching research content, which has the value of in-depth research and development. The construction of multi-modal teaching theory needs to be improved in the development process. In this research, several theoretical directions are put forward: in the multi-modal theoretical research, teachers and students should make full use of this mode, with the help of this model to further enhance the learning effect. Since the early 1990s, the related research has started to rise, and multi-modal theory has been widely used in other disciplines. The multi-modal teaching method is combined by information technology and multi-media technology. Starting from the students' perception, it integrates the students' quality resources in an all-round and comprehensive way, it realizes a kind of attribute transformation from mind learning to body learning. The most influential theories include the effective combination of multimodal theory with physiology and psychology.

There are great differences between traditional teaching mode and multi-modal research. The differences are shown in the following aspects: (1) the traditional mode belongs to the current mainstream mode structure, that is to say, teachers teach in class, the source of information and knowledge transfer is teachers, and the students are learners. (2) there are differences between information and knowledge transfer models. Teachers transfer knowledge and information effectively by means of language and other elements to form symbols. (3) under multi-modal conditions, it can save resources better and create a more powerful collective learning atmosphere, which has an important promoting role in effectively cultivating the concept of collectivism. The shortcomings of traditional teaching have also begun to show up in the long-term development process.

Teacher-led classroom model can not fully and effectively mobilize the enthusiasm of students, this model will more easily lead to a decline in students' desire to learn, therefore, can not better guide and train the student to form the creative thinking ability. Taking the unified teaching method will also result in teaching activities can not meet the characteristics of various types of students, can not reach the true sense of teaching according to students' aptitude. Not conducive to guide students to develop their own creativity. Of course, in the use of teaching materials and update also can not meet the actual needs. Multi-modal teaching can give full play to the advantages of teachers and students. On the one hand, teachers can effectively use the extended material to meet their own use of professor symbols. Students can also use more diverse ways of learning, can give full play to the function of this mode. For example, in the process of learning ancient poetry and prose, when teachers explain the artistic conception of poetry, the traditional teaching method is mainly to use language symbols to complete the description of the content, but with multimedia, students can be brought into a real situation by means of images and videos.

2 Teaching content of academic English writing for postgraduates

In the process of academic English writing, students need to show the research process in a systematic, orderly, efficient and well-grounded way through language expression, organization structure, grammatical rhetoric and argumentation strategies. Thesis writing based on literature reading is an important indicator of academic English writing ability and a necessary skill for learners to gain academic community recognition. It refers to the process in which the author applies the relevant texts to his own writing after quoting, rewriting, absorbing or expanding them, which includes three aspects: literature selection, literature reading and literature citation.

First of all, literature selection is "the basis of innovation ability of postgraduates". In the age of Internet and digital library, the types of document search engine emerge in endlessly. If learners do not have scientific literature search knowledge, they can not quickly select the latest literature or representative literature in a research field by "academic search engine" or "professional database", it is unable to carry out the later research work effectively. There are some problems in the topic selection of graduate thesis, such as insufficient theoretical depth, lack of innovation, lack of cutting-edge, and so on.

Second, the importance and characteristics of academic literature reading, as well as the development of academic literature reading ability, has been one of the important research topics in the applied linguistics. The process of reading literature is "the process of absorbing and criticizing the ideas of predecessors and stimulating self-thinking and seeking for writing inspiration". Literature reading "helps graduate students to understand and study research methods and academic norms, and has a profound impact on the follow-up study and research of graduate students.". In order to solve the problems of thinking activities of English majors in the process of English literature reading evaluation, such as lack of clarity, lack of organization, research ability and innovation ability, etc. , offering a course of literature reading and evaluation for postgraduates of foreign language institutes. Guide the students to read and evaluate the literature, at the same time, consciously cultivate their high-level thinking, and properly solve the problem of weak thinking ability of graduate students.

In addition, document citation is an important writing skill, which can not only show the breadth of the author's knowledge of the specific research, but also skillfully embed the author's research in the discourse of the subject. In recent years, domestic and foreign scholars through interviews, thinking experiments, text analysis, case studies and other methods, this paper makes an extensive and in-depth study on the relationship between the content and strategies of English Literature Citation and writing quality. Foreign language learners have different ideas about the purposes, contents and strategies of literature citation, and they do not know how to use appropriate strategies to avoid plagiarism. The effective literature citation strategy has the remarkable promotion function to the academic writing achievement.

3 The teaching strategy of multi-modal academic English writing for graduate students under the background of information technology

The emergence of multi-modal discourse analysis theory has become a hot topic for many scholars to study. Many scholars have applied the multi-modal theory to the teaching of reading, writing, audio-visual and speaking, the application and practicability of multi-modal theory are expounded by theoretical and empirical methods. The teaching means that the teacher guides before writing and the platform that the students evaluate each other are the important components of the whole dynamic development process. The emergence of multi-modal theory redefines the concept of writing teaching, and the process-based writing teaching is no longer confined to a single text form, but integrates various modes such as image, animation and sound to attract students' attention, to arouse students' enthusiasm and interest in writing, to realize the mutual cooperation between teachers and students, to broaden the dimension and depth of students' writing, and to improve students' writing ability.

Writing is the output of the learner after he internalizes the new knowledge and establishes some relations with the existing knowledge in the process of language learning, in the whole process of writing instruction, teachers should first consider how to input well so that students can accept the input in the most natural and full manner, students then combine their existing knowledge system output high-quality, reasonable layout, have their own views of the article. Paying attention to the pre-writing process, understanding and arranging the relevant materials before writing have a vital influence on the dimension and depth of students' thinking about the problem, and can make students generate interest and inspiration in autonomous learning, give play to the advantages of individual, so that students in the data search stage play their subjective initiative. Next, the group leader organizes the group members to discuss the difficulties and common problems on the BBS. The teachers should play the role of macro-management and solve the problems reflected by the students. Because of the amount of material collected before writing, the writing of the first draft will be easy. Teachers make clear the tenor, register and mode of the topic on the network platform, and use the openness, sharing and real-time interaction of BBS to solve students' common problems, that in itself is one of the advantages of a web platform. In the peer-review stage, students are encouraged to exchange their compositions with each other to understand the perspectives of other students on this topic, students attach great importance to the writing ability of their peers, and the accompanying peer pressure will improve their writing.

At this stage, the teacher should guide the students to evaluate their peers' compositions from different angles, not only in grammar and spelling, but also in structure, logic and argumentation, on this basis and then back from these points of view of their own essays, so that students in the process of understanding the meaning of "critical thinking". In the revision stage, the students themselves carry on the deep processing to their own composition, combine the teacher and the peer's feedback, pay attention to the vocabulary, grammar, sentence structure, the article layout, the thought dimension and the depth synthesis aspect. It is worth noting that the next step for teacher review is the key to the entire interactive model. First of all, the teacher's comprehensive level and thinking space can let students to their own strengths and weaknesses more clearly, the revision process more targeted. At this stage, teachers can summarize students' common problems, such as grammar, vocabulary selection and expression, and share them with others through links. In the final stage, the accuracy of the text is emphasized. The teacher can link the excellent composition to the students at this time, and the students can improve their writing and appreciation level when they appreciate and share the peer composition.

Based on the multi-modal theory and the concept of online interaction, the teacher-student interaction model of writing teaching network is constructed: in the pre-writing stage, teachers rely on the multi-modal teaching model to provide students with multiple modes besides language, by inputting multi-modal resources such as image, sound and video, students can understand and think about writing topics from multi-dimensional perspectives. At this time, the network platform also provides a large number of information input materials for teachers and students. In addition to searching and summarizing themselves, teachers also encourage students to search and summarize multi-modal network materials themselves, encourage students to think about writing topics from different angles and in different directions. Teachers can collect the data into a PPT, and then put it in the group sharing, so that students before writing immersed in the multi-modal information resources provided. In addition, to improve students' Understanding of multimodal theory, we should let them know that images, sounds and other non-linguistic modes play an important role in human communication. Before writing the first draft, the teacher explains the writing strategies, ideas and skills, as well as the flexible use of the text structure, syntactic structure and vocabulary transformation. In the next group discussion, let the students talk about their initial ideas, in the exchange of learning to appreciate and evaluate the views and ideas of others. Students submit their essays online after completing the first draft, and then conduct peer review in groups. Students are encouraged to use multi-modal resources to communicate with their peers and improve their multi-modal reading and construction abilities. In grouping, teachers should take into account the difference of students' level and ensure that students of all levels can get objective and correct evaluation. The content of the evaluation is submitted directly to each person's online submission, which is convenient for teachers to monitor the objectivity of the evaluation. At the same time, using the network platform to share students' writing resources, so that students can share each other's information and resources. The next step is for each student to revise his or her composition according to the comments and his or her new thinking, and submit it again to the network platform. Students hand in final draft after teacher's review. Teachers at all stages pay attention to students' network platform, and solve students' perplexities and bottlenecks in dealing with others and their compositions by means of online communication.

4 Conclusion

Writing teaching network teacher-student interaction model for the majority of teachers and students to provide a variety of writing teaching resources and network applications. The teaching of multi-modal writing breaks through the single modal form of text, and uses various resources such as images, sounds, colors, movements, eyes and spaces as writing resources to input students, the use of online platform provides resources for multi-modal resources, teacher-student interaction, and student-student interaction, in different stages of writing teaching, it provides space for teachers and students to input, appreciate, evaluate, correct and revise, and provides a platform for real-time communication and sharing of resources and achievements.

5 Acknowledgements

The project "cultivating intercultural communication-oriented postgraduate English teaching research" is funded by the 2021 of Hubei University of Arts and Sciences(Fund project number:XK2021019).

References

1. Wang Junju, Yang Kai, Sun Tianfeng. Research on the Literature Citation Characteristics of English Graduate Academic Writing [J]. Foreign Languages, 2017, (5): 56-64.
2. Li Zhi, Wu Zengxin, Shi Baohui. Research on the Reform of Academic English Teaching Mode for Postgraduates in Agriculture and Forestry Colleges [J]. Degree and Postgraduate Education, 2017, (4): 56-59.
3. Shao Hui. Research on the construction of English academic writing ability of graduate students in science and engineering colleges [J]. Journal of Beijing University of Technology (Social Science Edition), 2017, (2): 82-88.
4. Zhong Jiabao. Research on the pragmatic identity construction of postgraduate academic English in the new era [J]. China Foreign Languages, 2018, (5): 62-66.
5. Zheng Fei. Project-based teaching method for doctoral students' academic English writing stage [J]. Journal of Beihang University (Social Science Edition), 2021, (6): 133-139.

Accurate Monitoring Technology of Core Body Temperature of "yunwentie" Human Body Based on TRIZ Theory

QiuYi Yan, *Tongyao* Shen, *Shuilian* Yan, *Yinglu* Kuang, *Dingxiu* Fan, *Wenwen* lao, *Xiaqi Lan*

School of Nursing, Guilin Medical University, Guangxi, China 541001

Abstract. Novel coronavirus pneumonia is a new type of core body temperature measurement product. Methods using TRIZ theory, the problems of existing temperature measurement methods in China were analyzed. Using the corresponding invention principle, solve the relevant problems of thermometer in clinical use, and put forward innovative solutions. Results according to the innovative design idea of TRIZ theory, it was applied to the new human core temperature measurement product. Novel coronavirus pneumonia is difficult to achieve the current core temperature monitoring of new crown pneumonia epidemic. Conclusion the application of TRIZ theory to the research and development of new human core thermometer products defines the improvement direction of thermometer, and puts forward a feasible scheme for the design and development of new human core thermometer products.
Keywords: TRIZ, Accurate core temperature detection, The thermometer, Innovative design.

1 Introduction

At present, there are various thermometers in the domestic market, such as mercury thermometer, electronic thermometer, multifunctional infrared thermometer and so on. Mercury thermometer is the most common one. It is cheap and can be measured at any time without external power supply. It is deeply trusted by people, especially medical workers. However, the traditional mercury thermometer is not environmental friendly, difficult to read and unable to record data, which is inconvenient for patients with acute and serious diseases, the elderly, infants and so on [1]. The electronic thermometer displays the body temperature in digital form by using the determined relationship between the physical parameters of some substances (such as resistance, voltage, current, etc.) and the ambient temperature. The reading is clear and easy to carry. Its disadvantage is that the indication accuracy is affected by factors such as electronic components and battery power supply [2]. The multifunctional infrared thermometer can measure both ear temperature and frontal temperature. The dual function mode can be used in different situations. The defect is that the infrared thermometer is a non-contact thermometer, which is greatly affected by the environment. If there are obstacles such as hair or liquid at the measuring part, the measurement result deviation is too large [3]. The above existing temperature measurement methods can not quickly and accurately realize the real-time statistics and processing of group body temperature data. Based on the requirements of clinical and health monitoring, the operability and signal characteristics should be considered in the instrument design to simplify the system structure and portable design. The existing conditions can not achieve non-invasive body surface nondestructive measurement of core body temperature. The existing thermometer has large errors, which brings inconvenience to clinical research and disease diagnosis, treatment and nursing.

2 TRIZ theory

In the most popular sense, innovation is the process of creatively discovering and solving problems. The powerful role of TRIZ theory is that it provides systematic theoretical and methodological tools for people to creatively discover and solve problems.

The modern TRIZ theoretical system mainly includes the following aspects:

1. Innovative thinking methods and problem analysis methods
2. Evolution law of technological system
3. Technical contradiction resolution principle
4. Standard solution of innovation problem
5. Invention of problem solving algorithm Ariz
6. Knowledge base based on physics, chemistry, geometry and other engineering principles

After more than half a century of development, TRIZ theory has developed into a mature theory and method system to solve the practical problems of new product development. It has strong practicability and has been tested in practice. Its application field has also expanded from the field of engineering technology to management, society and so on. TRIZ theory has received great attention in western industrial countries, and the research and practice of TRIZ has been popularized and developed rapidly. Now it has achieved significant benefits for many well-known enterprises. Practice has proved that the use of TRIZ theory can greatly speed up the process of people's invention and get high-quality innovative products. It can help us systematically analyze the problem situation, quickly find the essence or contradiction of the problem, accurately determine the problem exploration direction, help us break through the thinking obstacles, break the thinking set, analyze the problem with new vision, carry out systematic thinking, predict the future development trend according to the law of technological evolution, and help us develop competitive new products. This paper will analyze the research and development of new human core temperature measurement products through innovative thinking methods and problem analysis methods, and determine the preliminary idea and final scheme by discovering the core problems and fundamental contradictions [4].

3 Analysis of TRIZ theory

3.1 Problem description

At present, the most common types of thermometers used in clinical diagnostic research and nursing work are mercury thermometer (Fig.1a), electronic thermometer (Fig. 1b), followed by multifunctional infrared thermometer (Fig. 1c).

Mercury thermometer: it is easy to break and may be polluted by mercury; The measurement time is relatively long, which is inconvenient for patients with acute and serious diseases, the elderly, infants and so on. Electronic thermometer: the indication accuracy is affected by factors such as electronic components and battery power supply (Fig. 1 D).Infrared thermometer: non-contact thermometer, which is greatly affected by the environment. If there are obstacles such as hair or liquid at the measuring part, it is easy to affect the measurement results (Fig. 1 E).

a

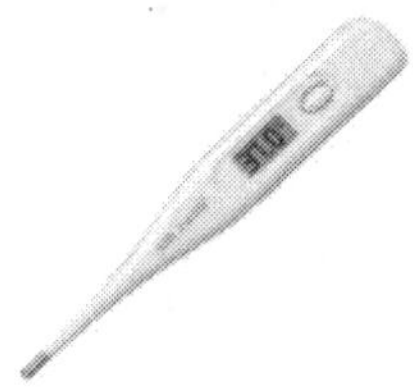

b

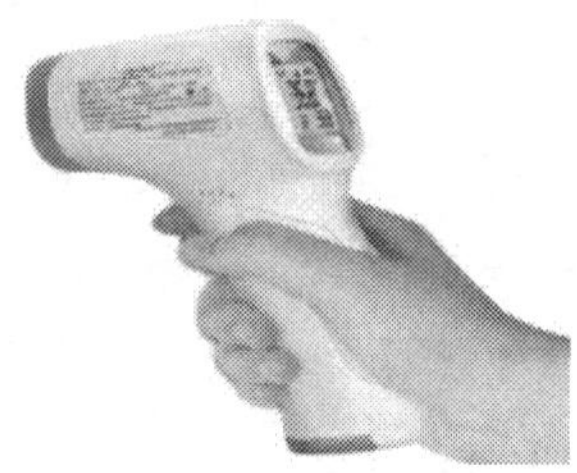

c

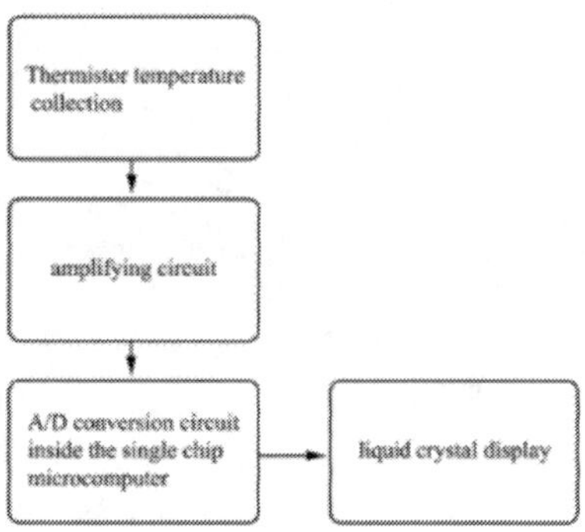

d thermometry principle of electronic thermometer

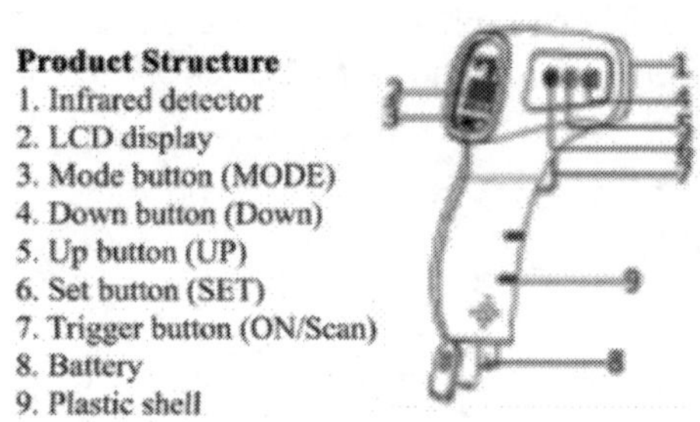

e Structure of infrared thermometer

Fig.1a mercury thermometer

Fig.1b electronic thermometer

Fig.1c multifunctional infrared thermometer

Fig1d thermometry principle of electronic thermometer

Fig 1e Structure of infrared thermometer

3.2 Problem analysis method

Problem analysis method is a systematic method to find out the problem and determine the cause of the problem according to the thinking process of solving the problem.

According to the thinking procedure for solving problems, the general steps of problem analysis are:

The first step is to identify the problem;

The second step is to further analyze the problem;

The third step is to explain what is deviation and what is not deviation, and ask questions;

The fourth step is to find out various factors that may lead to deviation;

The fifth step is to confirm what is the real cause of the deviation from the above factors, and check and verify it.

Following the thinking process of solving problems, the first step is to put forward problems; The second step is to clarify the problem stage; The third and fourth steps are the hypothesis stage; The fifth step is the hypothesis testing stage [5]. According to the problem description, the disadvantages of the current common thermometer are analyzed, and the figure shown in Figure 2 is obtained:

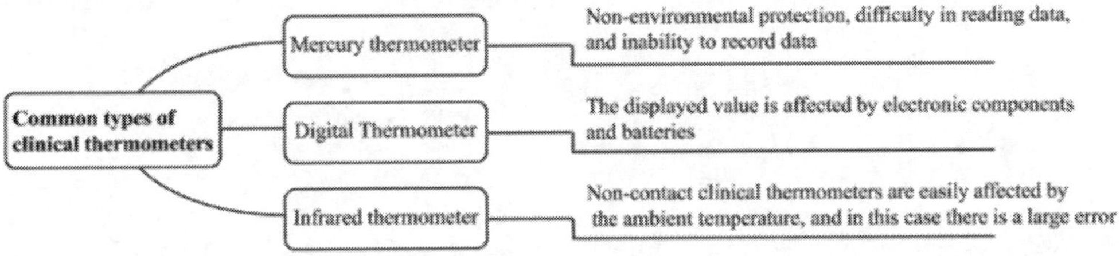

Fig. 2 Analysis of issues.

The problems are analyzed and discussed. ① ~ ③ s describe the problems existing in the existing thermometer, which can be solved by using TRIZ theory. As there is no more accurate and systematic temperature measurement method, this product is based on the internet medical background and combined with big data to develop a new human core temperature measurement product - body temperature patch. Therefore, the research and development of this product mainly solves the problems that the existing thermometer is not accurate enough and it is difficult to systematically and dynamically monitor the changes of core body temperature and basic vital signs of the human body, makes up for the shortcomings of the existing temperature measurement methods, simplifies the nursing procedures, and makes the measurement more accurate, efficient and convenient on the premise of safe measurement.

3.3 Analysis of system functional components

The portable "cloud temperature paste" thermometer has obtained the patent certificate and carried out the preliminary sample research and development. The design uses a high-precision medical digital semiconductor temperature sensor for data acquisition, and uses a Bluetooth chip to wirelessly transmit the data to the mobile phone app. The mobile phone app can store and record the body temperature monitoring data, so as to monitor the temperature of the measured personnel within a certain period of time. Through the temperature curve display of mobile app, the temperature change and development trend of the measured object can be observed, so as to monitor the health status of the measured object. Compared with the traditional mercury thermometer, the system has fast temperature measurement speed, the data can be recorded and reproduced, and a health database can be established according to different measured objects to record the body temperature in different time periods. At the same time, the temperature measurement accuracy can reach ± 0.1 °C, meeting the requirements of temperature monitoring accuracy [6], as shown in Figure 3.

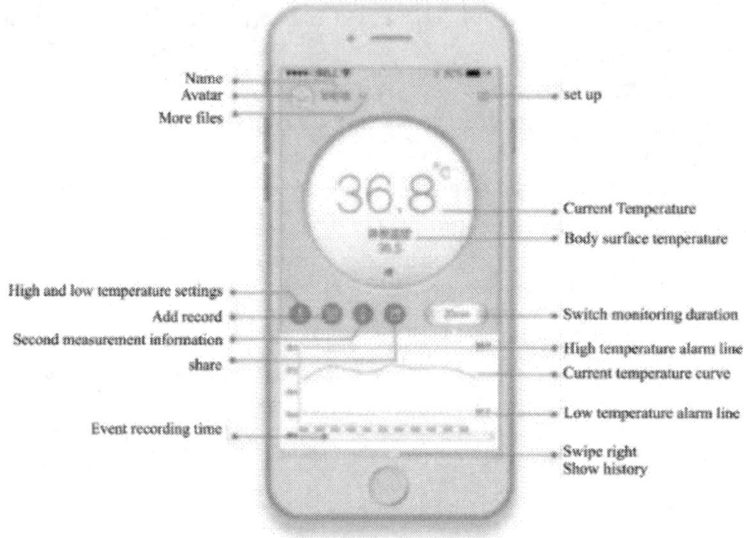

Fig.3.

4 TRIZ theory to solve the problem

4.1 Product introduction

The "cloud temperature patch" is a circular patch, which can separate or connect different external shapes. It is equipped with a shell protection to protect the internal technology chip of the product from damage. Combined with the developed big data platform, patients can measure themselves with only one patch. The measurement location is in the body surface projection area of pulmonary artery catheter. The monitoring is 24-hour continuous monitoring. Through the cloud platform, the data can be connected to mobile phones, community hospitals, doctors, etc., as shown in Figure 4.

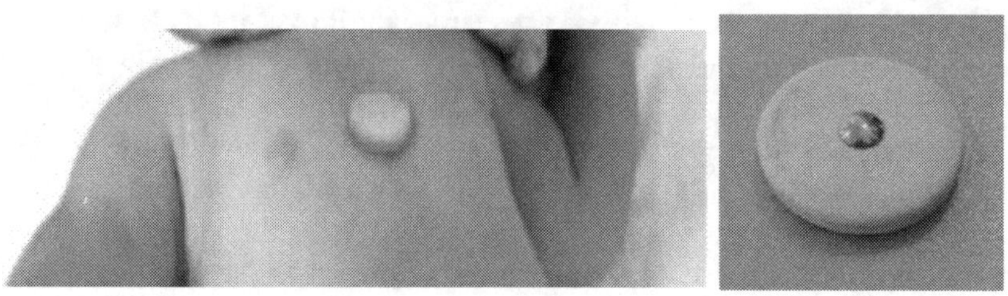

Fig.4.

4.2 Application of inventive principles

(1) It is easy to observe at any time. If the temperature changes, it will cause color changes. Measures can be taken immediately to avoid aggravation of the disease and prevent it from happening;

(2) The product is light, easy to carry and easy to install.

(3) Compared with other temperature measuring tools, it is more economical to measure the same number of temperature measuring points.

(4) Avoid scraping, scratching, folding or high temperature during use and storage. If you are not skilled in stripping, you can tear it gently from the edge. Don't pick it randomly to avoid damage. The temperature measuring part pasted shall be relatively clean and can be cleaned or wiped with alcohol or other solvents, so as to paste firmly [7].

4.3 Product instructions

① Separate the patch at the bottom of the product and place it on the chest of the human body for data measurement. After about 3-5 minutes, remove the patch;
② Paste the patch back to the bottom of the product, and directly read the body temperature on the electronic display screen of the product in real time;
③ Connect the product with Bluetooth devices such as mobile phones to obtain more real-time monitoring data;
④ After multiple measurements, the system will obtain the change curve of the measured data and the analysis and comparison results. When the body temperature is too low or too high, the alarm will be triggered.

4.4 Determination of innovation scheme

The achievement of the new precise monitoring technology of human core temperature is called "cloud temperature paste", as shown in Figure 5.

The trend of novel coronavirus pneumonia is accelerating. The vital signs such as body temperature, respiration and pulse are of great importance for epidemic prevention and control, case selection, timely diagnosis and timely isolation treatment. In the physiological range of fever, the accuracy of peripheral thermometer is uncertain. The existing temperature measurement methods can not quickly and accurately realize the real-time statistics and processing of group temperature data. In order to solve this problem, the research and development of "cloud temperature paste" provides a more convenient and efficient service means for temperature monitoring of susceptible people. Dynamically monitor the changes of vital signs and inflammatory indicators of suspected cases, and accurate temperature measurement will guide the diagnosis and management of high-risk groups. Clinical diagnosis should learn from the core temperature of human body, so as to solve the problem of screening epidemic patients from the perspective of accurate nursing technology and timely diagnosis and treatment.

The novel coronavirus pneumonia is a novel coronavirus pneumonia portable intelligent early warning device developed by combining the large data platform developed by the enterprise with the most vulnerable and vulnerable groups. It is a new portable intelligent early warning device for rapid detection of suspected cases. It can quickly and accurately monitor the vital signs and inflammatory markers of suspected cases, and screen, diagnose, and diagnose new crown pneumonia patients. The evaluation of treatment effect in the process of treatment after diagnosis plays an important role in clinical guidance.

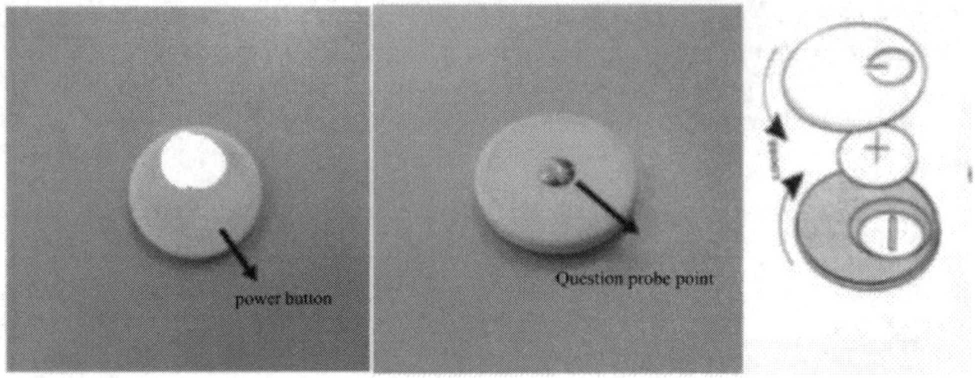

Fig. 5.

5. Discussion

TRIZ theory provides a series of innovative principles and methods to solve problems, and provides important ideas for the innovative design of body temperature paste. The innovative scheme solves the problems that the existing temperature measurement methods in China are insufficient in accuracy and safety, and can not meet the needs of clinical treatment and nursing. At the same time, it also improves the selectivity of temperature monitoring. It is suitable for various situations with simple and convenient operation and low complexity. It provides more accurate vital sign data and judgment factors for medical care, and reduces a series of diseases caused by the difference of body temperature. The patch does not use toxic chemicals such as mercury, which reduces the risk of mercury pollution. The innovative invention scheme has applied for a national invention patent.

References

1. https://baike.baidu.com/item/%E4%BD%93%E6%B8%A9%E8%AE%A1/1210134?fr=aladdin
2. https://baike.baidu.com/item/%E4%BD%93%E6%B8%A9%E8%AE%A1/1210134?fr=aladdin#7
3. https://baike.baidu.com/item/%E4%BD%93%E6%B8%A9%E8%AE%A1/1210134?fr=aladdin#7
   https://baike.baidu.com/item/TRIZ%E7%90%86%E8%AE%BA/936992?fr=aladdin
4. https://baike.baidu.com/item/%E9%97%AE%E9%A2%98%E5%88%86%E6%9E%90%E6%B3%95/2977944?fr=aladdin
5. https://www.tes-cn.com/tes-cn_Affiche_1135342.html

MATEC Web of Conferences **359**, 01004 (2022) https://doi.org/10.1051/matecconf/202235901004
ISC-FTE 2022

Application of Modern Information Technology in College Physical Education

Zhixiang Jin, *Wei* Zou

Huazhong Agricultural University, Wuhan, Hubei, 430070

Abstract. In the face of the reality of the work of traditional college physical education personnel, through the integration of modern information technology, not only helps the return of the value of sports aesthetics, but also helps students to participate in the process of sports practice, the profound perception of sports aesthetics in physical education helps students to form a comprehensive understanding of the value of multiple education in physical education. Based on this, in the intelligent education environment, based on modern information technology, innovation and reconstruction of college sports education mechanism, to truly achieve the goal of college sports education. This paper analyzes the background and main points of the application of modern information technology in college physical education, and puts forward the application strategies.
Keywords: Modern information technology, University, Physical Education, application strategy.

The integration of "Internet +" online teaching has become an important development direction of higher education in China, in the future, the traditional classroom teaching and the on-line teaching with modern information technology will coexist for a long time and merge deeply. In the context of comprehensive education, giving full play to the advantages of modern information technology will stimulate the enthusiasm of students to participate in sports learning, and cultivate and strengthen their sports spirit.

1 The application background of modern information technology in physical education in colleges and universities

The course reform of higher education towards the online resource sharing mode is the inevitable choice in the information age, the only way of physical education, and the important means to promote the better development of physical education. In the process of physical education in colleges and universities, students will be fully integrated into the work of physical education on the basis of optimizing the form of physical education.

1.1 Necessary means to meet the needs of students for different types of PE courses

After the application of "Internet + Education Model" in practice, the traditional teaching model needs to be developed to "Internet +" innovative teaching model to meet the needs of students for various types of special sports courses. Most colleges and universities offer physical education courses with limited teachers or equipment, and offer only a limited range of physical education courses. Some schools can not offer swimming, skating, golf and other physical education courses because of the conditions, which is in conflict with the students' desire for the study of different types of sports courses. It is of great significance to carry out the construction of on-line resource sharing model for physical education courses. On the one hand, through resource sharing, the limitation of physical education teachers and equipment resources in colleges and universities has been alleviated, on the other hand through the mode innovation to expand the students to learn sports knowledge and enhance sports skills channels.

1.2 Effective ways to broaden students' sports vision through multi-channels

Effective sharing of online educational resources can facilitate the development of in-class and out-of-class integration of physical education, and schools can explain technical actions in class to enable students to understand relevant knowledge, students can choose special VR and related APP learning when they review and preview. The on-line resource sharing model helps to expand students' sports vision through multiple channels, for example, teachers can influence students by updating sports news, Star News, focus events and other knowledge about sports through micro-blogs, wechat public accounts and sports forums, cultivating students sports cultural accomplishment. In addition, schools can also promote sports-related content through new media and expand the impact of sports education through retweets and likes, so that students can have more opportunities to learn about sports events, in order to arouse students' interest, the school can also organize sports-related knowledge competitions based on the Internet, and guide students to answer questions online, so that while effectively disseminating sports knowledge, strengthen the sense of belonging and pride of students. Schools should give full play to the advantages of the network, the use of new media to continue to publicize sports-related knowledge, so that the idea of lifelong sports rooted in students' hearts.

1.3 The inevitable trend of optimizing physical education curriculum resources

Physical Education curriculum information sharing is not only within the school, but also between different schools, even between international schools, each school has its own gold medal curriculum or excellent teachers, through the university physical education information sharing, excellent physical education teachers not only guide the students, but also benefit colleges and universities all over the country, greatly improving the quality and efficiency of physical education. The on-line resource sharing mode can also analyze students' sports learning situation through data sharing and big data analysis, so as to better discover the characteristics of college students' sports learning, finding out the problems in college students' PE learning and making corresponding changes in time can make college students more interested in PE courses and stimulate their enthusiasm in learning.

2 Application requirements of modern information technology in physical education in colleges and universities

2.1 Grasp the "vertical link" between the upper and lower sections

The most outstanding problem of physical education course is the lack of organic connection between the study stages, especially the lack of cohesion of the content. "the main line is to cultivate students' interests, develop their exercise habits, master sports skills and enhance students' physical fitness, improve the national physical education and health curriculum standards, and establish a link-up system for physical education courses in universities, middle schools and primary schools." It is mainly to determine which period of study what, reflect the cohesion, and full of logic, according to the content of the difficulty and the law of physical and mental development of students, such as progressive concrete embodiment. Only the top and bottom link, can avoid the low-level repetition, the student's physical education study can reflect in each stage study the suitability and the actual effect, thus satisfies the development demand.

2.2 "Horizontal consistency" in the same section

The evaluation can give a relatively accurate conclusion about the learning effect of physical education course. The research on the evaluation of physical education curriculum has already attracted the attention of scholars, and it focuses on the establishment of the evaluation system, and tends to the evaluation of physical education curriculum in colleges and universities, no one is interested in the related research that the PE curriculum evaluation reflects the "horizontal consistency" of the same study period. Physical education curriculum integration "horizontal consistency" refers to the students in the same class in the evaluation results, specifically refers to the level of sports ability to maintain relative consistency, including basic sports ability and specific sports ability, this "horizontal consistency" of the paragraph provisions, but also more conducive to the realization of the content of the link.

2.3 Maintain the "inner unity" of the target sports

The goal of physical education curriculum embodies the direction of the reform of physical education curriculum, and the relative unity of the goal should be maintained in the teaching process of special sports. At present, there are few researches on the goal orientation of "Inner Unity" in teaching special sports. The research on the integration of physical education curriculum puts forward that the goal of special sports should be to maintain "inner unity", which means that no matter which special sports, we should focus on the teaching of the four dimensions of "knowledge, ability, practice and health" and the overall promotion of students. For example, basketball should not only focus on the teaching of basketball-related knowledge, students should also emphasize the mastery and flexible use of basketball skills, the formation of basketball sports ability can not be ignored to develop good basketball habits, as well as the promotion of health. The reason for putting forward "inner unity" is to suggest that any special sport should not only be limited to imparting skills, but should be able to realize its functions in all directions and play an educational role in guiding and training students, in fact, it is through basketball to achieve a comprehensive educational function. This kind of unification request, can fully manifest the special movement the comprehensive education value.

2.4 Pay attention to "formal combination" of internal and external study

The effect of PE curriculum implementation determines the speed of curriculum reform, affects the depth of curriculum reform, and also reflects the breadth of curriculum reform.

The research on the implementation of physical education curriculum has always been a key issue of academic concern and discussion, most of the research focuses on the implementation of the status quo investigation, implementation strategy discussion, and some of the research to explore the effectiveness of implementation. The implementation of physical education curriculum involves a wide range and complexity, and how to effectively implement it is also one of the core issues of the integration of physical education curriculum, more systematic and more comprehensive to ensure the integration of physical education curriculum to comprehensively achieve the educational function of the discipline. The so-called "form Union" has the multi-dimension of the field. Only when the "form union" is realized at the level of implementation, the integration of physical education curriculum can have more outstanding effect.

3 The application strategy of modern information technology in college physical education

How to develop and construct the online teaching of physical education in colleges and universities in the new era lies in strengthening the deep integration of information technology and physical education teaching. As to how to achieve "deep integration", we should deeply understand the concrete contents of the reform of classroom teaching structure and implement the innovative teaching mode that can effectively change the classroom teaching structure, develop a wealth of learning resources for related disciplines.

3.1 Improving the understanding of online teaching of physical education in colleges and universities

In fact, more and more studies prove that online teaching is also suitable for physical education practice courses. With the coming of the information age, the pace of knowledge change and dissemination is quickening. Online teaching, with its advantages of no time and space limitation, convenience and rich resources, not only meets the diverse needs of students for sports knowledge, strengthening the body cognition of sports skills, improving the ability of self-study in sports, and widening the field of traditional offline sports teaching. However, we should be aware that the online teaching of physical education can not replace the offline teaching of physical education, but is a useful supplement and constant improvement of the traditional teaching of physical education. For the on-line teaching of physical education in colleges and universities, it should focus on how to realize the basic task of establishing morality and cultivating people, and how to adhere to the educational idea of health first, how to help students enjoy the fun of physical exercise, strengthen their physique, improve their personality and exercise their will, and then help students understand the basic elements of online teaching of physical education in colleges and universities, physical Education teachers should renew their teaching ideas, improve their information literacy, rationally design and apply online teaching, take students' physical and mental development and meet students' needs for physical education learning as the center, and play the leading role of physical education teachers, cultivating students' ability of self-study in physical education, strengthening their habit of physical training, building and sharing high-quality online teaching resources of Physical Education, building an online learning community of physical education, to achieve the physical learning interaction, physical learning records, physical examination and performance management and other links and elements of a comprehensive convergence and integration.

3.2 Construction of on-line teaching model suitable for college physical education

A deep understanding of the essence and connotation of online teaching of physical education in colleges and universities. The application of modern information technology in the field of education has given birth to on-line teaching. Its fundamental aim is to make use of information technology to better solve the problems existing in traditional teaching and thus improve the teaching quality, but its key point still lies in the deep integration of information technology and subject teaching. Setting up on-line teaching objectives of physical education in colleges and universities which accord with the functions and characteristics of on-line teaching. The functions and characteristics of online teaching mode of physical education in colleges and universities determine that there are obvious differences between its teaching objectives and offline teaching. Among them, the online teaching of sports theory can not only carry out the systematic dissemination of sports knowledge, but also integrate the rich cases vividly presented into the cultivation of patriotism, collectivism and socialist spirit, realizing the requirement of "ideological and political education" in physical education. On-line teaching of sports practice focuses on the cognition and understanding of sports skills, and the guidance and management of after-school physical training. In addition, on-line discussion, on-line homework, on-line test, but also through big data management to strengthen students sports awareness, sports habits and formation. Develop an online teaching plan for physical education in colleges and universities corresponding to the curriculum. The on-line teaching of physical education in colleges and universities is relatively independent and systematic, which involves not only the on-line teaching of a knowledge point or chapter, but also the on-line teaching of the whole course.

3.3 Improve the management and evaluation mechanism of online teaching of physical education in colleges and universities

The position and function of on-line teaching of physical education in colleges and universities. Since MOOCS appeared, online teaching has been widely welcomed by teachers and students all over the world, but it has also been questioned because of its own imperfections and the absence of the corresponding teaching management system. Many point of view that online teaching is only suitable for liberal arts courses teaching, physical activities are not conducive to the main type of practical sports courses. The on-line teaching of physical education can cultivate the students' ability of self-study, but it is also easy to form the situation of the students' self-indulgence in physical education. Therefore, on the one hand, the establishment of college physical education online teaching team. On the other hand, we should establish the perfect evaluation standard of P. E. Online teaching. We should not only consider the process evaluation of sports learning, such as the effective number of posts, teaching video learning time, after-school sports homework upload, but also reflect the summative evaluation. In addition, we should also give full play to the advantages of the internet big data to complete the dynamic management of Students' extracurricular sports learning and exercise. Choosing the appropriate online teaching method of physical education. The on-line direct broadcast teaching is not suitable for large-scale and normal physical education teaching because of its limitation. With the maturity of MOOC technology and online platform function, mooc-based online teaching has become the mainstream. At the same time, we should consider the difference between physical education theory teaching and physical education practice teaching, and whether the current online teaching environment is conducive to the achievement of physical education goals.

4 Conclusion

The development of on-line teaching of physical education in colleges and universities in the new era should be based on the needs of contemporary college students in their study and life, guided by the basic task of establishing morality and fostering people, and adhering to the first concept of health, and giving full play to the advantages of on-line teaching, improve the online teaching structure of physical education, enrich the online teaching resources system of physical education, and strengthen the deep integration with the traditional teaching methods of physical education, we should actively construct the school physical education teaching mode that integrates health knowledge, basic sports skills and special sports skills.

5 Acknowledgements

2020 Provincial Teaching Research Project of Hubei Province Colleges and universities, “A study on the training mechanism of sports core quality of top talents -- taking a talented class in a university as an example”; project number: 2020263.

References

1. Wang Guoliang. An empirical study on the introduction of flipped classroom into public physical education teaching in colleges and universities [J]. Journal of Xi'an Institute of Physical Education, 2019, 36(1): 110-116.
2. Liu Zhanlu, Feng Jiacheng, Yu Jiajun, et al. Research on the integration of MOOCs into sports education model classroom teaching [J]. Journal of Guangzhou Institute of Physical Education, 2019, 39(4): 110-112
3. Yuan Fang, Nie Yong. Research on the basis of ubiquitous learning and teaching support services in the age of intelligence [J]. Modern Educational Technology, 2019, 29(5): 26-32.
4. Mao Zhenming, Zhang Yuanyuan, Ye Ling. On the Lost and Return of Sports Fun in Physical Education Class [J]. Journal of Chengdu Institute of Physical Education, 2019, 45(2): 33-37+31-32+2.
5. Guo Shuai, Feng Xiaoli. Teaching and Learning of College Physical Education in the Age of Information Technology: Concept and Evidence [J]. Journal of Physical Education, 2021, 28(1): 108-113.

Research on Multi-interactive Teaching of College English under the Environment of Information Technology

Qiongjing Zheng

Nanning University, Guangxi, Nanning, 530200, China

Abstract. Multi-interactive teaching model is an important way to improve the quality of college English teaching, and it is also the key foundation to improve students' understanding of college English teaching content. In the increasingly mature environment of information technology application, students' learning enthusiasm will be fully stimulated, multi-interactive communication will be deepened, and students' English core literacy will be effectively enhanced by giving full play to the advantages of information technology, to upgrade the intelligence of college English teaching system. Based on the analysis of the multi-interactive teaching mode of college English under the information technology environment, this paper puts forward some practical strategies for innovating the multi-interactive teaching of college English by using information technology.
Keywords: Information technology environment; College English; multi-interaction; teaching research.

The multi-interactive teaching model of college English based on modern information technology can fully embody the teaching structure of "teachers leading students", and further improve the quality of English teaching by giving full play to the advantages of information technology, give full play to the role of teachers in teaching guidance and promotion, effectively improve students' ability to use English.

1 Multi-Interactive teaching mode of college English under the environment of Information Technology

Under the environment of information technology, the advantages of the multi-interactive teaching mode of college English are mainly reflected in the creation of a pleasant learning environment and the provision of rich teaching resources for students, guide students to actively participate in learning activities, students make full use of advanced teaching equipment and teaching resources, under the guidance and help of teachers, actively explore and participate in learning activities, to be the active constructor of knowledge meaning and the sharing of learning resources.

1.1 Teacher-Student interaction

First of all, the teacher-student interaction model is embodied in the face-to-face interaction in the classroom. Teachers can make full use of modern information technology such as network and multimedia, design pre-class tasks according to the actual level of students, and aim at different teaching contents, encourage and guide students to use the advantages of the network teaching environment to find relevant information. Classroom teachers use the network, video, audio materials, three-dimensional presentation of teaching content, through the five steps of task introduction, knowledge learning, scenario simulation and guidance, expansion and application, evaluation and detection, classroom teaching is carried out according to the teaching goal and the actual situation of students, according to different teaching contents, different teaching methods are adopted to make students participate in classroom activities to the maximum, so that students really become the main body of the classroom. In the process of teaching, we should adjust the teaching methods and strategies, arrange the teaching process reasonably, control the classroom teaching rhythm, help and guide the students in time, so as to activate the students' cognitive needs and stimulate their learning interest, encourage the students' cooperative consciousness and realize the cognition and externalization of the rules of language use. Secondly, the teacher-student interaction model also reflects the online interaction after class, students can ask questions through the teaching platform, teachers receive and carefully record students' feedback, for question-answering, evaluation and guidance. In order to solve the problems in the process of autonomous learning in time, remove their fear of learning English psychological barriers, enhance students' confidence in learning English, cultivate their ability to learn English on their own.

1.2 Interactive teaching mode

Compared with classroom teaching, the modern information technology teaching environment with the network multimedia technology as the core can provide students with massive and rich forms of learning resources, which rely on teaching software and teaching platform, the effective combination of classroom teaching and extra-curricular extension in the form of electronic teaching plans, electronic courseware, audio-visual materials and other teaching resources has transcended the limitation of classroom time and space, to enable students to choose learning materials that suit their needs through the internet and computer at any time and any place, to actively explore learning, and to experience a new three-dimensional teaching content, enjoy the text, sound, pictures, animation, film brought sensory stimulation, personalized learning. Online learning includes listening bar, spoken English show, English reading room, English learning club and online corpus. In the self-testing module, students can take self-stage tests to adjust their learning progress, teachers record the time and progress of Students' learning, and test the effect of students' learning. In the module of answering questions, teachers timely release the course assignments and answer the questions raised by students. This teaching platform provides students with rich learning resources and personalized learning environment, extends students' English learning space, and gives full play to students' learning potential.

1.3 Student-Student interaction teaching model

According to Constructivist theory, students' construction of knowledge not only depends on their own knowledge level and experience, but also depends on the joint discussion and understanding of problems among learning partners. Pre-class teachers put forward questions according to the teaching content and teaching objectives, design tasks, divide students at different levels into different study groups, carry out task division, cooperate with each other to consult materials, discuss and explore related issues; In class, we will have interactive communication through pairwork, group work, role-playing, situation simulation, problem discussion and topic debate After class, students use English corner, self-access Learning Center, social practice, forum, email, QQ and special study website to communicate and exchange information, cooperative learning, mutual help and common progress in a real environment. While improving the ability of using language, the students' ability of inquiry is strengthened, and the students' team spirit and ability of getting along with others are cultivated.

2 The characteristics of multi-interactive college English teaching in the information technology environment

With the support of modern information technology, the multi-interactive teaching mode of college English advocates the teaching idea of taking students as the main body and teachers as the leading part, which not only promotes the construction of teaching resources, it also provides flexible learning strategies. The multi-element interactive teaching model of college English integrates English classroom teaching with extracurricular practice and online learning effectively, to make English teaching toward personalized learning, autonomous learning, research-based learning direction of healthy development. The multi-element interactive teaching mode fully pays attention to students' subjectivity, realizes the teaching mode of teacher-student interaction, human-computer interaction, and student-student interaction, and highlights students' individualized learning, it has strengthened the students' autonomous learning ability and the ability of using language.

Based on the modern information technology, the multi-interactive college English teaching model fully reflects the teacher-led, student-oriented teaching philosophy, embodies the people-oriented teaching philosophy. This model attaches great importance to students' subjective status and respects students' needs and choices. It not only creates a relaxed and pleasant learning environment for students, but also provides them with real and rich learning resources, to encourage students to actively explore, learn independently and cooperate with each other, to promote the organic combination of learning and use, to maximize students' full participation in learning activities, and to tap students' intrinsic learning potential, develop their ability to find, analyze, and solve problems. The practicality of language determines that language learning can not be separated from imitation and practice. Modern information technology provides extensive learning resources and real context. Students can listen to lectures, open online classes, it can also make use of the real corpus provided by network teaching resources, multimedia courseware, English learning website and online corpus to make up its own information gap. Through inductive conversation model, teachers can create situational simulation to help students improve their language competence in real-life experience and cooperative interaction.

Cooperative learning is divided into teacher-student cooperation and student cooperation, classroom teaching, extra-curricular practice, network self-exploration of all kinds of learning are inseparable from cooperative learning. Teachers should design the content suitable for students' cooperative learning according to the teaching content, guide students' Cooperative Learning, supervise and coordinate the learning process, take the initiative to discuss problems with students, communicate effectively and solve problems. Students cooperate with each other through mutual assistance, competition, motivation, sharing and other ways to carry out cooperative learning.

The evaluation of college English courses under the modern information environment adopts the evaluation system which combines formative evaluation with summative evaluation, aiming at strengthening the monitoring and evaluation of students' learning process, and strengthening the cultivation of students' learning habits, encourage students to participate effectively in the process of autonomous learning. Formative assessment includes classroom performance, work display, second classroom practice and web-based learning records. The evaluation methods can be divided into self-evaluation of learning, mutual evaluation of students, group evaluation, teacher evaluation and so on. In summative evaluation, the oral test is introduced, which is combined with the written test and the test paper is selected randomly from the test paper library.

3 The multi-interactive teaching strategy of college English under the environment of Information Technology

Under the environment of information technology, when carrying out the multi-interactive teaching of college English, we should strengthen the combination of classroom teaching, online and offline teaching, and this kind of mixed teaching mode is more targeted, teachers should make full use of the advantages of information technology and introduce high-quality learning resources into the curriculum through the application of various information-based teaching platforms and resources, facilitate students to quickly and easily access to knowledge, expand the scope of English learning. Teachers should also recognize that classroom learning time is relatively limited and that quality resources can be pushed to students, who can use mobile devices to learn relevant resources and improve their English learning.

In the process of teaching, English teachers can make full use of information equipment to carry out multi-interactive English teaching. Teachers can choose English content suitable for students as the theme, let students through the Internet search related video resources to learn English each other. College English teachers can let students take video clips, choose the characters in the video in free groups, and use audio recording software to imitate the plot and the communication between the characters, through the audio submitted by the students, the teacher analyzes the problems existing in the spoken English of the college students, and points out the mistakes in the application of the spoken English in a timely manner through the online communication, so as to guide the students to strengthen the related training. Through this kind of interaction, English teachers can not only deepen their understanding of college students, but also practice their oral English ability in a way that students like, so as to improve their English application ability.

It can be divided into four parts: listening, speaking, reading and writing, which can be integrated into input and output. Listening and reading are input skills, speaking and writing are output skills. To improve the ability of students to speak should start from the source, as much as possible to provide English language environment. In the classroom more oral practice and tests to encourage students to speak. To provide a variety of channels to expand students' English input channels. Teachers should take an active part in, guide and promote the activities of "English corner" on campus and the broadcasting of English radio, and organize various extra-curricular activities such as English speeches and English competitions, optimize the English learning environment on campus, so that students have more exposure to English, in order to make up for the lack of English classroom listening and speaking.

Network and multimedia technology provide material guarantee for the realization of classroom space-time extension, and build a platform for students to study independently. Set up an online learning system such as an online English course platform, so that students can study independently and complete exercises related to the text at any time, students are encouraged to collect cultural background knowledge, allusions, audio-visual images of the original version of English, and a database of training questions and mock-up questions in

CET4 and CET6 online for easy reference. Establishment of multimedia college English teaching model. In teaching, teachers should give full play to the advantages of rich experience and experience in classroom teaching, and can adjust teaching methods and contents at any time according to students' expressions and reactions, taking advantage of the advantages of multi-media teaching, rich resources on the internet and unlimited learning time and space, the combination of multi-media and network is introduced. Let students not only feel intuitive classroom teaching, after class and enjoy the online lectures, online discussion opportunities, independent choice of their own learning methods and learning materials, and at any time to express their views.

Set up an online question answering system. In the process of learning, we can further improve the network learning system and realize the network answering questions, so that when students have questions, they can contact teachers by e-mail and other forms, and can get help online consultation. Establish a reasonable evaluation system. Teaching assessment can be divided into formative assessment and summative assessment. Formative assessment, which usually includes attendance, classroom performance, assignments, quizzes, mid-term exams, online self-access records, etc. Summative assessment refers to the final course examination and various level examinations. Both summative assessment and formative assessment should be emphasized in the college English testing and assessment system, as well as the assessment of students' comprehensive quality and development ability.

4 Conclusion

In the process of college English teaching, the adoption of multi-interactive teaching methods can provide more opportunities for college students to display, through multi-interaction, college students can find out the problems and shortcomings in the process of English learning in time, and make clear their own advantages in English learning. It is also helpful for English teachers to deepen their understanding of college students, in order to grasp college students' individuality and English learning needs, strengthen students' ability of learning and applying English, and realize effective English teaching.

5 Acknowledgements

A study on the quality administration of blended teaching in English curricula in private higher learning institutions (No. 2021ZJY737), a project supported by Guangxi Educational Science Planning Foundation in 2021

References

1. Zhu Chengqi, Yu Shuhao. Exploration on the Multi-Interactive College English Teaching Mode [J]. Education and Teaching Forum, 2018, (50). 173-174.
2. Guo Juan. Research on multi-interaction college English teaching mode from the perspective of "Internet +" [J]. Comparative Research on Cultural Innovation, 2018, (31). 98-99.
3. Pan Peng. Analysis of the changing roles of college English teachers and students in the multi-interactive teaching mode under the Internet background [J]. English Square, 2018, (1): 69-70.
4. Ma Jing. Analysis of the Multi-Interactive College English Teaching Mode [J]. Overseas English, 2019, (18): 117-118.

5. Wang Cunbao. Analysis of the multi-interactive college English teaching model [J]. Farm Staff, 2019, (5): 140+147.

Study on the Comprehensive Management Mode of Diabetic Patients under the Mode of Medical Care Combined with Old-age Care

Xiaodong Xing, *Wenyan* Wang

Inner Mongolia Minzu University, Inner Mongolia, Tongliao, 028000, China

Abstract. Diabetes is a chronic disease with a high incidence rate. In the context of health care in China, an intelligent monitoring system combining medical care with old-age care has been established based on the service model of combining medical care with old-age care, in view of the reality of diabetes patients, while accurately and effectively identifying the nursing problems of patients with diabetes, a scientific and reasonable individualized nursing plan will be formulated, and the relatively limited nursing resources will be fully utilized, to improve the quality of nursing care for the aged with diabetes mellitus. Based on the model of medical care combined with old-age care, combined with diabetes patients, a comprehensive management model was put forward.

Keywords: Combination of medical care and nursing care, old-age service mode, diabetes patients, comprehensive management mode.

The combination of medical care and nursing care is a kind of old-age service with Chinese characteristics. With the rapid development of population ageing and aging in China, the combination of medical care and nursing care has emerged as the times require. The combination of medical care and nursing care is a combination of medical care institutions and old-age care institutions. Old-age care resources are integrated into medical care resources, the new service mode of medical care, nursing care, health care and nursing care is to meet the needs of the elderly and to provide continuous care for them.

1 Background of the research on the service mode of the combination of medical care and old-age care

At present, the rapid development of the population ageing not only brings heavy burden to the society and the family, but also puts forward higher request to the social medical level and service. China's elderly population has a large base, a rapid growth and a significant aging trend, coupled with changes in family structure and large population mobility, which has led to a significant increase in the demand for elderly care and medical services, in particular, the number of patients with subacute stage, end-stage patients, respiratory care needs are numerous. In 2017, WHO issued guidelines on integrated care for the elderly in the community and guidelines on assessment and intervention pathways for integrated care for the elderly in the community. In April 2019, the State Council General Office issued an opinion on advancing the development of old age services, which pointed out that promoting the high quality development of old age services, including enhancing the capacity of medical and Nursing Combined Services, we will promote the integrated development of old-age care at home, in communities and in institutions.

Medical care combined with old-age service mode is compared with the traditional mode of old-age care. It is a fine tradition of the Chinese nation to provide for the aged. At present, there are several types of medical-care combination models in China: first, medical institutions are set up in old-age institutions; second, old-age institutions are set up in hospitals; Fourth, the medical institutions to the old-age and medical services to combine the direction of transformation; fifth, the community home old-age service institutions and medical institutions to establish a cooperative mechanism. These models have their own advantages and disadvantages, so it is necessary for the builders of medical institutions and medical workers of the elderly to explore and innovate in practice.

The new "combination of medical care and old-age care" model is based on the above-mentioned traditional old-age care model, with the help of various resultant forces such as institutional mechanisms, participation forces, scientific and technological support, etc., with the participation of professional nursing team, general practitioners, rehabilitation doctors and medical institutions, we will provide professional medical services for the disabled, the mentally retarded and the elderly to improve their quality of life, in order to achieve the "disease treatment, non-disease health care" of the old-age model. In the new model, it specifically includes the contents of "medicine", "health care", "maintenance" and "Nursing", "medical" refers to medical services, health advisory services, health examination services, disease diagnosis and treatment services, "health" refers to serious illness rehabilitation services; "Yang" refers to life care services, mental and psychological services, cultural activities services, etc.; "Nursing" refers to nursing services, palliative care services, etc. . Prevention and Treatment Combined Services, actively promote community medical care and care for the aged. Combined with the construction of embedded community service facilities for the elderly, the joint establishment of Community Health Service stations, with medical care services as the main function, to provide medical services such as seeking medical care and dispensing, chronic disease management and family sickbeds for the care-giving elderly, and to carry out targeted health services combining prevention and treatment for the elderly in the community, promoting the organic combination of community medical care service and old-age care service. Family doctors sign contracts to extend community geriatric health services. The new service mode of "combining medical care with nursing care" takes "medical care + rehabilitation + Nursing + old-age care" as the main line, and carries out targeted care and health management to the residents according to the level of need nursing, a model for the provision of convenient and effective comprehensive services that integrate life care, cultural entertainment, health care, spiritual comfort, emergency relief and palliative care for the elderly.

2 The challenge of diabetes management in China

Diabetes self-care is challenging, and education alone is not enough to help people with diabetes build confidence, motivation, and skills for self-management, a combination of program education, counseling, and skills training is required to enable patients to acquire and maintain self-management skills, including practical problem-solving skills and coping skills, over a period of time, then repeat the training courses to improve their self-management skills and self-efficacy.

2.1 Complications of diabetes

The high complication rate of diabetes in our country causes great damage to the health of the residents, and causes heavy burden of disease to the society, second only to cardiovascular disease and tumor. The diabetes patient suffers from the cardiovascular disease probability to be higher than non-diabetes 2-3 times, the diabetes research institute result indicated that the diabetes patient because of overweight brings the high blood fat is the important reason which produces the cardiovascular disease; People with end stage renal disease are 10 times more likely to have diabetes than those with early stage renal disease, and diabetic retinopathy affects a third of people with diabetes, the leading cause of vision loss in adults. Complications due to poor blood sugar control are an important cause of disability in patients with type 2 diabetes. The Glycated hemoglobin is the gold standard for blood sugar control, according to a large randomized trial of diabetes complication control (DCCT) in the United States, strict control of blood glucose close to the normal level can effectively control the occurrence and development of diabetic complications.

2.2 Treatment of diabetes

As a metabolic disease, hyperglycemia is seen as the tip of the iceberg in metabolic disorders, and patients with type 2 diabetes mellitus (T2DM) are often associated with other metabolic syndromes, such as hypertension, dyslipidemia, obesity, and so on, these combined diseases make T2DM complications of the risk, speed of development and harm significantly increased. For people with Type 1 diabetes, a healthy lifestyle can be achieved through proper daily insulin treatment, regular monitoring of blood sugar, a healthy diet and a regular lifestyle, delay or avoid many of the complications associated with diabetes. While a healthy lifestyle is the cornerstone of treatment for type 2 diabetes, including a healthy diet, increased physical activity, smoking cessation programs and maintenance of a healthy physical quality, so far, the cause of type 2 diabetes is not fully understood and can not be cured, requiring patients to follow medical advice throughout their lives. Every year, about one-third of the diabetic patients in our country develop gradually and produce all kinds of acute and chronic complications due to lack of basic knowledge of diabetes, failure to take active treatment measures or improper treatment, such patients usually have a low level of disease-related knowledge, poor self-management ability, poor compliance with medication, affect blood sugar control effect, need to be supplemented with good health education.

3 The comprehensive management mode of diabetic patients under the mode of medical care combined with old-age care

The elderly patients are the mainstream of diabetes. Because of the increasingly prominent problems of the elderly, the transformation of the old-age service institutions and the integration of various resources has become a general trend. As a result of the huge change of people's life style and diet structure as well as the increasing life pressure and other factors, more and more people are suffering from diabetes, the incidence of diabetes remains high. In order to further improve the elderly diabetic care, we need to actively integrate into the medical care model.

3.1 Continuum of care model

The integrated management model of diabetes patients under the mode of medical care combined with old-age care practice the concept of holistic medical care and continuous care, the technology of discharge preparation service (the core technology of continuous medical service) is applied to improve the quality of medical service and facilitate the smooth transition from hospital to community and back to family. The medical care ward carries out the evaluation of the discharge preparation of the admitted patients, selects the high-risk patients, provides the discharge preparation service before and after the discharge of the patients, and establishes the working procedure of discharge preparation service, the contents include: Comprehensive Evaluation of patients (post-admission, pre-discharge and post-discharge follow-up) ; multi-disciplinary team participation in the preparation of discharge plan; full communication with patients and their families to improve the quality of care during hospitalization; Health education and caregiver training were conducted to guide the self-care knowledge and skills needed after discharge from the hospital, and seamless handover was completed in cooperation with the contracted doctors in the community health center, so that the community medical care extended to home care.

3.2 Family-based participatory tele-care model

With the development of network technology, remote management based on network has become the trend of continuous nursing. But at present, most of type 2 diabetes patients are middle-aged and old, their education is not high and their hearing and vision are generally decreased. The supervision and support given by family members in diet, medication and exercise can greatly promote the rehabilitation of patients with diabetes. The family participation type long-distance nursing model is the union family support and the network long-distance nursing merit, may fully enhance the diabetes patient's self condition management ability.

3.3 The health belief model

Health belief model is a theoretical model to explain health-related behaviors by using social psychological method. From the perspective of HMB, people's behavior is determined by their beliefs about disease and behavior, that is, when people feel threatened by disease, the benefits and barriers of accepting behavior are likely to lead to behavior. Until now, integrating diabetes management self-efficacy into HBM. The older the patients were, the worse their self-management behavior, and the strongest predictor of self-management behavior was self-management efficacy, suggesting that the patients with high self-management efficacy had better self-management behavior.

3.4 Multi-disciplinary joint diagnosis and treatment, all-round management model

The general management clinic for Type 1 diabetes includes a combination of diabetes specialists, diabetes educators, nutrition, exercise rehabilitation, ophthalmology and psychology. The diabetes specialist is responsible for assessing blood sugar control, assessing growth and development, adjusting treatment regimens, and answering questions. The nutrition division is responsible for calculating the daily energy intake, which is converted into the intake of various foods; conducting dietary surveys and assessments for three consecutive days; and conducting dietary education on diabetes and addressing the problems in the diet of patients, dietary and nutritional advice is given and patients are provided with individualized and nutritious meals. The rehabilitation department is responsible for exercise assessment, cardiopulmonary function assessment, assessment of bone and joint conditions, exercise prescription. The psychological department is responsible for the evaluation of the emotional state of the patients and their family members, and the evaluation of the cognitive function of the patients. Ophthalmology is responsible for diabetic retinopathy screening. The diabetes educator is responsible for the establishment of patient files, long-term follow-up, evaluation of treatment results, evaluation of patients' self-management ability, individualized education based on the evaluation results; Check the insulin injection site and so on.

3.5 Diabetes support group model

The effectiveness study of the group of diabetic friends was mainly based on the intervention study of Educational Support Group. Based on the support group of patients, the intervention of self-management plan was designed, and the effect of self-management plan was followed up after 12 weeks and 24 weeks respectively, improving diet knowledge and skills, exercise and self-monitoring behavior can improve group leader's ability, and improve patients' quality of life in general. In addition, an intervention study was conducted to design a self-management program for type 2 diabetic patients, the results showed that there were significant differences in HbA1c, fasting blood glucose, body mass, diabetes cognition and health belief between the participants and non-participants. The self-efficacy and depression of the patients can be improved by using the support group model of the diabetic patients and the intervention of the support group, hbA1c, cholesterol level and quality of life were improved.

4 Conclusion

At present, diabetes is prevalent in the whole world, and it has become one of the main diseases that cause death and disability. Under the guidance of the mode of combining medical care with old-age care, through the integration of prevention and medical and health resources, to establish and improve a working mechanism and service system for the prevention and treatment of diabetes mellitus with clearly defined responsibilities, orderly connection, cooperation and interaction between disease prevention and control institutions and medical and health institutions, actively establish a data management system for diabetes patients, build a "medical prevention and integration" to manage diabetes in the whole process, implement a "prevention-intervention-treatment" comprehensive system, and explore an intelligent management model for diabetes, improve the health status of patients with diabetes.

5 Acknowledgements

Inner Mongolia Natural Science Foundation, Project No. 2018LH08050

References

1. Quan Xuhong, Jiang Mingjun, He Yan. Application and effect of home-based medical care combined model in elderly patients with type 2 diabetes [J]. Medicine Forum, 2017, (2). 11-13.
2. Wang Hongyan, Lin Qin, Zheng Yuanyuan, et al. Research on the application of the "combination of medical care and nursing care" mode based on the Internet of Things in elderly type 2 diabetes [J]. Chinese Rehabilitation Medicine, 2018, (1).6-8.
3. Hu Xiujing, Wu Xiaoya, Wang Jiaji, et al. Review of the development of the

integration of medical care and health care and health management in my country from the perspective of chronic disease prevention and treatment [J]. Chronic Disease Prevention and Control in China, 2019, (8).561-564.

4. Fang Hong, Yao Hong, Liu Yinan, et al. Health status of the elderly and coping strategies for general practice services under the mode of combining medical care and nursing care [J]. Chinese General Medicine, 2020, (S2): 20-22.
5. Wei Wen, Tang Xiaobo, Sun Shaoqing, et al. Effects of comprehensive health management model on self-management ability and HbA1c, FPG and 2 h PG levels in elderly hypertensive patients with diabetes [J]. Chinese Journal of Gerontology, 2021, (10):2186-2189.

The Construction of Chinese Culture in English Teaching from the Perspective of the Internet of Things

Wei Zhao

Dongbei University of Finance and Economics, Liaoning China, 116025

Abstract. At present, the country pays more and more attention to English teaching(ET). In the process of ET, Chinese culture(CC) is the basis for students to learn English, and it is also an important way to cultivate cross-cultural communication skills and comprehensively use modern language to conduct thinking activities and solve practical problems. However, the reality is that ET is facing more and more serious cultural deficiencies. English learners are constantly exposed to Western culture and gradually weaken their Chinese cultural identity. Therefore, it is very necessary and urgent to carry out the construction of CC in ET. At the same time, the Internet of Things is a product of the emergence and development of the information society. It can not only realize the sharing and utilization of information resources, but also meet the learning needs of students through various intelligent terminal devices, thereby further expanding the scope of application in ET and improving the level of educational informatization. Therefore, it becomes more and more meaningful to study the construction of CC in ET in the context of the development of the Internet of Things. This article uses questionnaire surveys and data analysis methods to understand people's views on the introduction of CC into ET through questionnaire surveys, so as to further complete the work of constructing CC in it. According to the survey results, among the 120 interviewees, 46 people think it is necessary to infiltrate CC in English learning, while only 7 people think it is unnecessary. It can be seen that, in the eyes of most of the interviewees, it is necessary to understand and learn CC in ET, and it will also be of great help to English learning.
Keywords: Internet of Things, English Teaching, Chinese Culture, Cultural Construction

1 Introduction

In the context of globalization, China, as a big country, is inextricably linked with the world. And ET is an indispensable part of Chinese education. However, there are many problems in ET, such as the lack of CC. Under the new situation, how to effectively use and develop university resources to improve the quality of talent training and to construct CC has become a topic of current concern. At the same time, with the continuous development of Internet of Things technology, the education field is also actively exploring new models, while also providing new ideas for the reconstruction of CC.

At present, the research results on ET and the introduction of CC are very rich. For example, Shi Zijuan proposed a college ET model based on the perspective of the Internet of Things. Through sufficient speech perception and information transmission, students can improve their ability to use language and achieve the goal of college ET reform in the new stage [1]. Liu Yingxin believes that it is necessary and feasible to maintain cultural safety in college ET, and it is necessary to introduce and learn CC under the guidance of reasonable principles [2]. Zhang Shenghua introduced the phenomenon of the lack of CC in college ET from a cross-cultural perspective, and emphasized the need to enhance students' cultural recognition ability and improve students' ability to express CC in English [3]. Therefore, this article combines the perspective of the Internet of Things to explore the construction of CC in ET from a new perspective, which has important research value and educational significance.

This article mainly discusses these aspects. First, the Internet of Things and its related content are explained. Then, it also introduces the current research of ET and the necessity of constructing CC in ET. Finally, a questionnaire survey was carried out around the question of "whether it is necessary to infiltrate CC in English learning", and the survey results and analysis conclusions were drawn.

2 The construction of Chinese culture in English teaching from the perspective of the internet of things

2.1 Internet of things

Certain achievements have been made in the development of the Internet of Things technology, and it has gradually penetrated into the education field. It came into being as a new thing and idea. The concept of the Internet of Things was first proposed by Kevin Ashton. It uses wireless sensor networks, global positioning systems, radio frequency identification systems, two-dimensional codes and other equipment and information collection systems to allow various objects to go online in accordance with the agreed protocol, and realize intelligent perception and communication through information sharing and communication. The core of the Internet of Things is the Internet, but the main part of the Internet is people and computers. The Internet of Things can realize the communication between things and things and between people and things [4-5]. The architecture of the Internet of Things is shown in Figure 1.

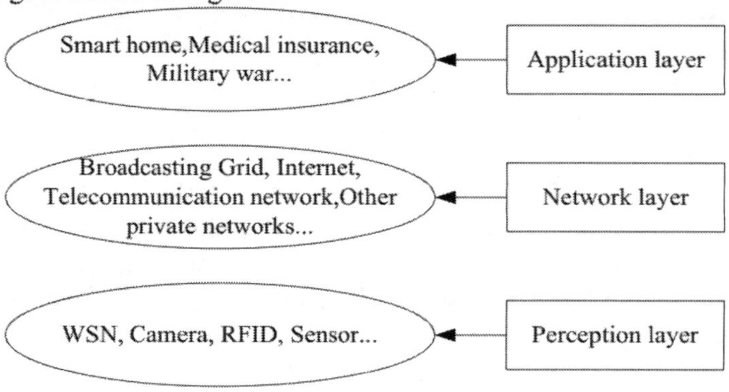

Fig.1. The architecture of the Internet of Things.

The core of the Internet of Things is depth perception, extensive networking and intelligent processing. Among them, the perception and communication capabilities of objects play a fundamental role in the application and development of the Internet of Things. It combines traditional sensors with network communication technology and is one of the key implementation and support technologies at the end of the Internet of Things.

With the continuous development of the Internet of Things technology, the education field is also actively exploring new models and has made certain achievements. For example, sharing teaching resources through network platforms such as "campus network" and "micro classroom". At the same time, a variety of forms such as the interconnection mechanism between the school and the enterprise, the school-enterprise cooperation relationship, and the student's personal learning space have been established to realize the exchange of information between teachers and students, and the development of online course testing activities to enrich the ET system. The scope of functions and application scenarios of the sci-fi, promote teachers and students to communicate with each other in practice and improve their mutual understanding ability, collaboration level and comprehensive quality [6-7].

2.2 The relationship between language learning and culture

Language is the carrier of culture, and language and culture are closely connected. As we all know, learning English is inseparable from the language environment, but second language acquisition cannot replicate the language and culture in the process of language acquisition. Chinese-English learners pay more attention to excellent teaching materials, foreign teacher training and practical opportunities for overseas exchanges. Everyone often forgets that the most common language environment for Chinese and English learners is the native language environment.

From the perspective of the influence of culture on language. Culture initially restricted the formation and development of language systems. For example, the formation of lexical meaning is inevitably restricted by various cultural factors, including local ethnic traditions, customs, religious beliefs, lifestyles, and even geographical environment. By learning any language vocabulary, you will find that the relationships between all the objective things mentioned and abstract concepts and their names permeate the actual experience and experience, and all reflect the understanding and feelings of real people. Secondly, culture also affects the way people think and express themselves in language. Some researchers believe that language use and language use analysis should be combined with the appropriate social and cultural environment, social origin and speaker's status, and specific communication situations. The influence of culture on language can be said to be very far-reaching. It includes not only the obvious elements of language, but also the internal thinking of language and the external expression of people. Pragmatics is the main branch of linguistics, which emphasizes the study of meaning in context. Every social custom includes all aspects of its own culture [8-9].

In summary, it can be said that learning CC can cultivate students' deep understanding of language, and also provide them with a lot of real and fresh language materials for learning English, so as to help students learn English better. Language and culture live in harmony, interdependence, influence each other, and are inseparable. ET should fully consider the cultural function of the language, not simply use language as a symbol, but also make full use of the role of mother tongue culture in understanding the target language and culture to help improve students' overall language ability, thereby enhancing students' communication between different cultures Awareness and ability.

2.3 The necessity of constructing Chinese culture in English teaching

The core of cultural construction is a relationship of mutual influence, interaction and restriction between people and the environment. In this process, we need to use a certain form to express. This change is mainly reflected in the individual's attitude towards things or social phenomena and their behavior patterns. To a certain extent, people will make appropriate adjustments when they are influenced by the outside world to change their behavior patterns or conceptual patterns, so that the social and cultural construction can be improved and improved.

The construction of culture is a complex process. It is not only affected by factors such as social development, historical evolution and realistic conditions, but also restricted by factors such as national traditional habits and customs. First of all, culture has a certain degree of stability, because it is not a product of immutability and immutability. Secondly, what is involved in the construction is the formation of stable social relations after the integration of nations, regions and countries. Finally, there are differences and diversity among different ethnic groups. Therefore, only through the analysis and research of language and cultural environment can we better realize the goal of Chinese cultural construction [10-11].

In the process of learning, English learners are constantly exposed to Western culture and gradually weaken their Chinese cultural identity. The role of CC in ET cannot be ignored. It can not only help students understand Western countries and their mother tongue, but also improve learners' understanding of China and CC. At the same time, it is necessary to actively use this new situation to provide more opportunities for foreign language teachers to help students solve certain problems and cultivate their understanding of traditional CC. A thorough understanding of mother tongue culture is a gradual process. In this process, only by consciously comparing CC and Western culture can one gradually understand the other's culture in English learning, identify with one's own culture, and overcome cultural shock [12].

3 Questionnaire and research

3.1 Questionnaire design process

In this questionnaire survey, the subjects selected for the survey are English teachers, scholars, English majors and non-English majors. The questionnaire survey was conducted in N City, with a total of 120 respondents. Through the issuance of online questionnaires or paper questionnaires, the collection and quantitative analysis of the information filled in by users are carried out to draw conclusions of the questionnaire.

(1) In the preliminary preparation of the questionnaire, the number of questions should be as concise as possible to avoid fatigue of the interviewees.

(2) The questionnaire is released. Questionnaires were distributed through online questionnaires, on-site questionnaires, and inviting friends to help ask friends and students around them to fill in the questionnaires. A total of 120 questionnaires were distributed and 120 valid questionnaires were returned. The questionnaire recovery rate was 100%.

(3) Questionnaire analysis. Organize the collected questionnaire information to get the required information data. Analyze the results of the questionnaire, including the respondents' opinions on whether it is necessary to infiltrate CC in ET. Some of the results obtained from the questionnaire are as follows.

3.2 Calculation method of reliability in questionnaire survey

In the questionnaire survey, whether it is an online questionnaire or an on-site questionnaire, there may be more or less errors. Among them, random errors are unavoidable, and systematic errors should be avoided or reduced as much as possible. Because in the actual questionnaire survey, the systematic error is difficult to decompose. Reliability is generally defined as the proportion of the variance of the true score in the total variance. The specific calculation method is shown in formula (1)(2)(3).

$$Y = R + A + F \tag{1}$$

$$\vartheta_Y^2 = \vartheta_R^2 + \vartheta_F^2 \tag{2}$$

$$Conf\left(Questionnaire\right) = \sqrt{\vartheta_R^2 / \vartheta_Y^2} \tag{3}$$

Among them, Y is the data obtained by directly conducting the questionnaire survey, R is the real statistical data, A is the error data in the questionnaire survey system, and F is the random error data. ϑ_Y^2 is the variance of the measured data, ϑ_R^2 is the variance of the true statistical data, and ϑ_F^2 is the variance of the random error data.

3.3 Questionnaire survey content

The first part is the selection of 120 citizens of City N. The interviewees are divided into English teachers, scholars, English majors, and non-English majors according to their occupations, and investigate the necessity of infiltrating CC in English learning.

The second part is to sort out the information collected in the questionnaire and understand the necessity of 120 interviewees to infiltrate CC in ET. Part of the questionnaire survey results are as follows.

4 Analysis and discussion

In this questionnaire, the necessity of permeating CC in ET was conducted on English teachers, scholars, English majors and non-English majors. The survey results are shown in Table 1.

Table 1. Results of the questionnaire.

Respondent	Very necessary	Necessary	General	Unnecessary
English teacher	12	13	4	1
Scholar	11	15	3	1
English majors	10	13	5	2
Non-English majors	13	10	3	3

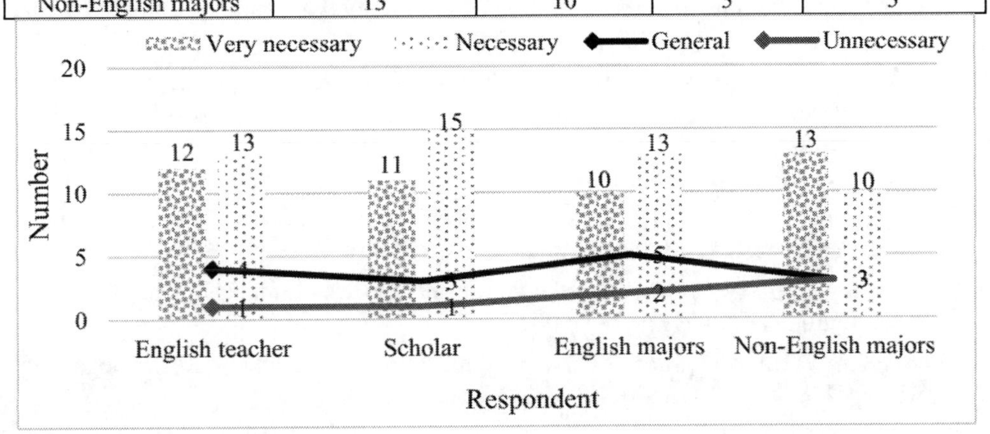

Fig.2. Results of the questionnaire.

It can be seen from Figure 2 that among the 120 interviewees, 46 people think it is necessary to infiltrate CC in English learning, while only 7 people think it is unnecessary. It can be seen that, in the eyes of most of the interviewees, it is necessary to understand and learn CC in ET, and it will also be of great help to English learning.

5 Conclusion

Nowadays, in the process of ET, English learners are constantly exposed to Western culture and gradually weaken their own Chinese cultural identity, thus neglecting the learning of CC. Understanding the connotation and background of CC is of great help to English learning. At the same time, as a product of the development of a modern information society, the Internet of Things can effectively integrate various network information resources to achieve common progress between people and the environment. At the same time, the use of Internet of Things technology to innovate teaching models is also one of the important conditions required for a country to improve the level of education. Therefore, this article has its value in studying the construction of CC in ET in the context of the development of the Internet of Things.

References

1. Shi Zijuan. The construction of a student-centered college English teaching model from the perspective of the Internet of Things[J]. 2021(2016-4):131-132.
2. Liu Yingxin. Research on the Introduction of Chinese Culture in College English Teaching from the Perspective of Cultural Security[J]. 2021(2018-24):36-37.
3. Zhang Shenghua. A review of the research on the phenomenon of Chinese cultural deficiency in English teaching from a cross-cultural perspective [J]. 2021(2018-6): 37-39.
4. Shi Yuan. An analysis of Chinese cultural aphasia in college English teaching from a cross-cultural perspective[J]. 2021(2016-1):76-77.
5. Wang Yanxiang. Some Thoughts on College English Classroom Teaching in the "Internet+" Era[J]. 2021(2020-19):213-214.
6. Zhao Ronggai, Zhang Jingtao. Research on the application of big data collection and analysis in the teaching of English writing in higher vocational colleges[J]. 2021(2016-30):23-23.
7. Wang Xuan. The reform and exploration of the interactive mode of English classroom teaching with big data[J]. 2021(2020-19):39-40.
8. Zhou Yan. The reform of college English information teaching in private colleges under the background of \"Internet+\"[J]. Journal of Hubei Correspondence University, 2019, 032(004):164-165.
9. Yang Xiyun. Reflection on Listening Classroom Teaching for Non-English Majors in Independent Colleges [J]. Curriculum Education Research: Foreign Language Learning and Teaching Method Research, 2018, 000(032): P.31-31.
10. Guo Xiaoying. Analysis of Chinese cultural imbalance in high school English teaching and its solutions[J]. 2021(2016-3):49-55.
11. Zhong Jing. The penetration of Chinese cultural identity education in college English teaching[J]. Nongjia Staff, 2020, No.663(15):219-219.

12. Guo Baoyun. The strategy of introducing Chinese culture into English teaching in higher vocational colleges[J]. Journal of Hubei Correspondence University, 2020, 033(003):156-157.

Study on Detection Technology of Antibiotic Content in Soil

Yi Jian[1], *Jing* Wei[2], *Jia* Hou[2], *Rui* Li[2], *Lianhui* Chen[3*]

[1]Department of Physics, North Sichuan Medical College, Sichuan, Nanchong 637100, China
[2]Department of Optometry, Grade 2020 Class 1, North Sichuan Medical College, Sichuan, Nanchong 637100, China
[3]Department of Chemistry, North Sichuan Medical College, Sichuan, Nanchong 637100, China
[*]Corresponding author: Lianhui Chen

Abstract. Since the discovery of antibiotics, has been widely used in the treatment of epidemic diseases, affected by this, a large number of antibiotics into the environment of the media, affecting environmental safety. Therefore, it is very important to detect the content of antibiotics in soil accurately and scientifically. At present, microbial assay, immunological analysis and physical and chemical assay are widely used in the determination of antibiotics in soil. The source and ecological risk of antibiotics in soil are studied, the concrete application of the technology is discussed systematically.
Keywords: Soil, antibiotics, content determination, detection technology.

The problem of antibiotic residue in soil has been paid more and more attention, and the detection technology of antibiotic content has been paid more and more attention. Currently, there are no limits on the emission and quality of antibiotics in environmental media, nor are they included in the Environmental monitoring and environmental management system, and there is a lack of technical specifications for the comprehensive utilization of livestock manure. Therefore, the contamination of soil environment by fecal antibiotics will become more and more serious.

1. Sources of antibiotics in soil and ecological risk

The use of veterinary antibiotics and medical antibiotics, as well as the sewage of antibiotic producing enterprises, are the three sources of antibiotics in the environment, among which, the use of veterinary antibiotics has exceeded 52% of the total, become the main source of antibiotics in environmental media. In recent years, the proportion of veterinary antibiotics production has been increasing year by year. In 2013, the production of 36 common antibiotics was 92,700 tons, of which 84.3% of veterinary antibiotics were produced, global meat production will rise by 30 per cent, with concomitant increases in antibiotic production and consumption. Because of the low absorption rate of veterinary antibiotics, 30% ~ 90% of them enter into the soil in the form of the excrement and urine of livestock and poultry, which results in the high residual level of antibiotics in the soil around the farm. Manure runoff is one of the main causes of soil antibiotic pollution in China, six times greater than in the UK and much of Northern Europe. The residues in soil with pig manure could be as high as 1000μg/kg, and the antibiotic content in soil without manure was 2 orders of magnitude lower than that in soil with manure.

Drugs and personal care products (PPCPs) , including human and veterinary drugs for treating diseases, personal care products such as bathing and skin care, and daily chemicals such as aromatherapy and disinfectants, are emerging pollutants with potential hazards to the ecological environment and human health. Their environmental presence, risk assessment and toxicology have attracted extensive attention in the environmental science field. The sources of PPCPs in the environment are very extensive. There are a lot of PPCPs in waste water and solid waste from pharmaceutical factory, hospital, livestock and poultry farm, and human daily life. As a class of PPCPs, antibiotics are widely used in animal husbandry, aquaculture and other fields. Most of the antibiotics enter the soil with the excrement, which is easy to cause soil antibiotic pollution. Because of its antibacterial property, the antibiotics entering the soil environment can only be removed by biodegradation, hydrolysis, photolysis or absorption by the roots of plants. However, the remaining trace antibiotics can cause certain toxicity and inhibition to microorganisms, animals and plants and soil ecology.

The ecotoxicity of antibiotics in environmental media was mostly quantified by half effect concentration (EC50) . The EC50 of Ciprofloxacin (CIP) and Ofloxacin (ofloxacin) ranged from 1.00 × 10-3 to 2.00 × 10-3 mg/l, the EC50 of OTC and TC were 1.04 and 3.31 mg/L, respectively. The EC50 of sulfamethoxazine, Trimethoprim and Florfenicol were 19.52,16 mg/L and 1.80 × 104 mg/L, respectively. Antibiotic residues in soil can enter human body through food chain as well as affect soil ecological environment. The daily exposure of Chinese residents to antibiotics through aquatic products was 2.0 × 10-7 ~ 2.7 × 10-3 μg/kg, while the absorption rate of FF in drinking water was low, about 64% was excreted through urine, and the daily exposure of children was 1.90 × 10-4 ~ 2.20 × 10-4 μg/kg. Some antibiotics (FQs, TCS) can accumulate slowly in human body at low dose, cause organ damage and abnormal function, or cause allergic reaction, hormone-like action, teratogenic, carcinogenic and mutagenic action, and increase drug resistance of pathogenic bacteria.

2 Detection technology of antibiotic content in soil

In the current study, antibiotics are usually analyzed by ultrasonic extraction, solid-phase extraction and liquid chromatography-tandem mass spectrometry (SPE-LC-MS/MS), which are optimized based on EPA 1694. Microbial diversity mainly refers to the study of soil microbial species and relative abundance, that is, microbial community structure. Soil is rich in microorganisms, the main microorganisms include bacteria, fungi, archaea, eukaryotes and so on. For the detection of different types of microorganisms, microbial diversity and related testing is divided into the following testing services:

2.1 Amplification subsequencing detection

Amplification sequencing is mainly used to detect a large class of soil microorganisms, such as bacteria or eukaryotes (16S/18S/ITS, etc.) , the species and relative abundance of a certain kind of microorganism, i. e. the community structure of microorganism, were obtained. With the popularization and deepening of the microbial diversity project, many scientific research projects now put forward higher requirements, hoping to detect the absolute number of microorganisms. At present, fluorescence quantitative PCR can be used for the quantitative detection of total bacteria or a certain class of microorganisms in soil. The combination of second-generation high-throughput sequencing and real-time quantitative PCR can be used to understand the species composition of microorganisms in samples as well as to obtain absolute quantification of specific species.

2.2 Metagenomic detection

Metagenomic detection is the use of the second generation of high-throughput sequencing technology to detect all soil microorganisms, including bacteria, fungi, archaea, eukaryotes and so on, the species, relative abundance and function of all microorganisms were obtained.

2.3 Functional gene sequencing

The second-generation high-throughput sequencing technique was used to detect the species and relative abundance of soil microorganisms with certain functions, such as nitrogen fixation. Soil is a major player in the Roger Fulford, and many of the microbes in the soil function as drivers of chemical cycles, participating in specific carbon, nitrogen, phosphorus, and sulfur cycles. A specific function of microorganisms is determined by a specific functional gene. By detecting a specific functional gene in soil, we can find the species and relative abundance of microorganisms with a specific function in soil.

2.4 Quantitative detection of microorganisms

It is primarily used to measure the absolute quantity of a certain type or specific microorganism. By using fluorescence quantitative PCR technique, not only a large class of microorganisms can be obtained, but also the absolute quantity per unit volume or weight of a particular microorganism can be obtained. This is a good supplement to the second generation high-flux technology, which can only be obtained in species and relative abundance. The combination of the second generation high-throughput sequencing and real-time quantitative PCR can be used to understand the species composition of microorganisms in the samples as well as to obtain absolute quantification of specific species.

2.5 Quantification of antibiotic resistance genes

The common antibiotic resistance genes are tetracycline resistance genes, sulfonamides resistance genes, erythromycin resistance genes, streptomycin resistance genes, etc., the absolute quantification of antibiotic resistance genes can be achieved by fluorescence quantitative PCR or high-throughput qPCR.

3 Conclusion and outlook

With the discovery of ARGs in all kinds of environments, ARGs can migrate and transform in all kinds of environmental media, and it is possible that the resistance gene can be transferred into the food chain, which makes the ARGS pollution global. The existing inspection regulations and most laboratories use the typical internal standard method to analyze several or more kinds of targets. There are structural and property differences between the internal standard and the target, the difference of loss, chromatographic retention behavior and response intensity in the pretreatment process leads to higher detection limit and poorer stability of the analytical method, which can not be used for the accurate quantification of low concentration antibiotics.

Blank test, each batch of samples to do at least 1 laboratory blank, the blank value of the measured elements should not exceed the lower limit of the method. If more than the reason must be found, re-analysis until qualified samples can be analyzed. Calibration curves shall be drawn for each batch of samples and the correlation coefficient of calibration curves shall be greater than or equal to 0.995. Calibration checks shall be performed for each analysis of 50 samples using a calibration curve intermediate concentration standard solution. The relative deviation between the determination result and the concentration of the last calibration curve should be less than or equal to 10% . Otherwise, redraw the curve. Accuracy Control, each batch of samples should contain certified reference material, and the accuracy should be controlled by adding standard to the samples within the given uncertainty, the recovery rate should be in the range of 80%-120% .

Nowadays, biosensors, optical sensors and electrochemical sensors based on CDs have been widely used in the detection of antibiotics in environment and food matrix. The small size effect and amphiphilic characteristics of CDs make it easy to be combined with chromatography for the detection of antibiotics, this may be due to the complex substrate for detection of antibiotics, the limited number of CDs available for antibiotic pretreatment, and the fact that the mechanism of CDS as chromatographic separation material is not completely clear. The substrate for the detection of antibiotics based on CDs materials is relatively single, and the analysis and detection of antibiotics in lake water, soil and other complex environmental samples are relatively few, because the size and function of CDs are easily affected by the temperature, time and the level of the experimenter, the synthesis of CDs with small size and special function is still in the laboratory stage. Some antibiotics have the characteristics of poor stability. When using the internal standard method to determine them, the storage time of the samples is short, and the requirements for analysis and determination are severe, try To finish pretreatment within 3 days and instrument analysis within 45 days.

SRGS and INTL1 were enriched by single and combined contamination of antibiotics and heavy metals, and the higher the concentration of antibiotics, the higher the relative abundance, the results showed that the resistance genes had common resistance mechanism to antibiotics and heavy metals. Tetracycline resistance Gene Tet A. Tet G. Tet X did not increase significantly, indicating that low concentrations of antibiotics and heavy metal contamination had no significant effect on TRGs induction. A representative soil sample was selected for high-throughput sequencing to study the microbial community structure under different initial addition amount. Actinomycetes and Proteus were the dominant bacteria in all the samples, 50.6%-70.4% . The relative abundance of Proteobacteria was 26.1%-41.3% , which was significantly higher than that of the control. The relative abundance of chloro flexi increased with the increase of the concentration of antibiotics, and the proportion of chloro flexi in the combined pollution of antibiotics and heavy metals was obviously lower than that in the single pollution, the results showed that the combined pollution of antibiotics and heavy metals could inhibit chloro flexi.

Antibiotics is a global pollution problem, which is directly related to the health and development of human beings. The pollution of soil antibiotics is closely related to the production and life of human beings. There are many ways for antibiotics to enter the soil. To control the pollution problem, we should further Quantitative analysis the forms of antibiotics in the soil and study the physical and chemical properties of the soil in the treated area, to find out the influence of soil environment on the transformation law of the forms of antibiotics in soil, so as to take effective measures to intervene and control the rate of the migration and transformation of antibiotics in soil, both economic and environmental benefits should be taken into account. We can also take measures to control the source and actively explore new repair techniques.

4 Conclusion

In summary, antibiotics have high toxicity and high ecological risk, soil as an environmental carrier of antibiotic enrichment, after the change of environmental conditions, antibiotics in soil can be released into water environment through surface runoff, soil runoff, underground runoff and so on to accelerate the diffusion and increase human exposure. Therefore, it is very important to do a good job in the detection of antibiotics in soil.

5 Acknowledgements

Science and Technology Project of Nanchong City, Sichuan Province; project name: Application of RLS in the determination of antibiotics in the surrounding environment of farmers; subject number:18SXHZ0282.

References

1. Nan Qiong; Tang Jingchun; Hu Yucheng; et al. Research progress of antibiotic detection methods in different environmental media [J]. Chemical Research and Application, 2017,(11):1609-1621.
2. Jin Lu; Chen Pinpin; Wang Xiao; et al. Research progress on pretreatment and detection methods of antibiotics in environmental samples [J]. Journal of Tianjin Agricultural University, 2017, (3): 92-96.
3. Zeng Qiaoyun; Ding Dan; Tan Xiao. Research progress on the pollution status and sources of tetracycline antibiotics in agricultural soils in China [J]. Journal of Ecology and Environment, 2018, (9): 1774-1782.
4. Ruan Linlin; Lin Hui; Ma Junwei; et al. Degradation and microbial response of single and compound antibiotics in soil [J]. China Environmental Science, 2018, (3): 1081-1089.
5. Qi Heming; Yan Hua; Zhang Lei; et al. Research progress of antibiotic detection technology in crops [J]. Journal of Food Safety and Quality Inspection, 2019, (15): 5098-5103.

Development of E-commerce Strategic Transformation under the Background of the Internet

Jie Cheng

Wuhan institute of Shipbuilding Technology, Hubei Wuhan, 430050, China

Abstract. At the end of the 20th century, the rapid development of e-commerce has become a new growth point for China's economy. The entire e-commerce industry is in a thriving state. The Internet and electronics are gradually changing people's lives. The Internet continues to develop in China, and the concept of "Internet +" has been integrated and penetrated into all walks of life in China. Various technical means and the "Internet +" policy have further promoted the development of e-commerce at a faster speed. In the past two years, the competition among e-commerce companies has become increasingly fierce, and e-commerce companies have undergone transformation, but the effect is not significant, and there is still a certain gap between expectations. Internet technology can drive the transformation of e-commerce to a certain extent. Therefore, this article is mainly based on the development research of e-commerce ST (strategic transformation) under the background of the Internet. First, a comparative analysis of the status quo of several mainstream domestic e-commerce companies shows that the domestic e-commerce industry has great potential for development. Alibaba's annual turnover is US$114.5 million, ranking first in the 2020 e-commerce transaction list. Secondly, it analyzes the effect of e-commerce ST under the background of the Internet. The results show that the rapid development of the Internet provides an excellent logistics system for the e-commerce industry, which helps e-commerce companies achieve differentiated and diversified development strategies.

Keywords: Internet, E-commerce, Strategic Transformation, Development Strategy.

1 Introduction

As the Internet began to enter the Chinese market at the end of the twentieth century, a long road of development began. E-commerce mainly uses Internet technology to directly sell products and services to consumers. The emergence of e-commerce has accelerated the pace of transactions in the business process, expanded the participants of e-commerce, and increased the proportion of direct sales from manufacturers to consumers. Therefore, this article mainly studies the ST of e-commerce under the background of the Internet, which promotes the transformation and development of the e-commerce industry.

There are many researches on the ST of e-commerce in the context of the Internet. For example, Huang P believes that "Internet +" is becoming more and more intense, and various industries are facing new market development situations, strengthen the use of Internet platforms, seize opportunities in advance, and complete the transformation of enterprises under the power of the Internet Development[1]. Jit SM pointed out that big data will start a major change in the era. With the development of e-commerce platforms, big data has played a huge role in the development of e-commerce [2]. Liu Y discussed the opportunities and challenges of e-commerce development in the era of big data, and its applications in various fields [3]. At present, although there are not a few researches on the ST of e-commerce under the background of the Internet, the number of studies on how to apply the Internet to the transformation of e-commerce is very limited. Therefore, this article mainly studies how the e-commerce industry can achieve transformation under the background of the Internet. It can accelerate the transformation of the e-commerce industry.

This article mainly studies the research on the ST of e-commerce under the background of the Internet. This article first studies the theoretical basis related to the Internet and e-commerce, and combines the results of the questionnaire to conduct the following analysis: first, a comparative analysis of the status quo of several mainstream e-commerce companies at home and abroad. The results show that compared with foreign e-commerce companies, the domestic market has greater development potential. Alibaba has an annual turnover of 11.45 million US dollars, ranking first in the 2020 e-commerce transaction list. Secondly, using a questionnaire survey experiment to analyze the effect of e-commerce ST in the context of the Internet, the results show that the rapid development of the Internet is conducive to the diversification and differentiation of e-commerce enterprises' ST.

2 Research on the development of e-commerce strategic transformation under the background of the internet

2.1 Internet technology

The concept of the Internet has been integrated and permeated into all walks of life in China. It has had an overwhelming impact in the past ten years. Because of the higher connection efficiency of the Internet and the connection of more people, money, and things, it has made the business rules of the entire world have changed[4]. The essence of the Internet is connection. The Internet has the characteristics of decentralization, de-productization, de-intermediation, and breaking information asymmetry. Before the advent of the Internet, the channels for humans to obtain information were single, complex, and inaccurate. In many cases, the transmission of information is ineffective and fruitless. After the advent of the Internet, information acquisition has become simple, making it possible to spread and transmit information at a high speed[5]. The advent of the Internet has made the input and output of information extremely efficient and accurate. The Internet has lowered various thresholds and promoted equality and sharing of information. In the era of sharing economy, the support of the Internet is behind it, such as e-commerce, which greatly reduces the cost of ordinary people 's online business transactions, and can expand their products to the entire domestic market through the Internet, and lower the threshold. Another example is Zhihu. All knowledge seekers can find their own knowledge and ask questions on the platform, which realizes the equality and sharing of information. These are difficult to achieve in traditional industries, and the cost of input is a threshold [6].

2.2 The Impact of Internet technology on e-commerce

E-commerce is a systematic use of information network technology as a means to engage in various network business activities centered on commodity exchange without geographical restrictions[7]. In the field of e-commerce, the biggest data must come from the purchase behavior of users. At present, there are already e-commerce platforms that have thousands of people based on users' buying habits, that is, users with different consumption characteristics may have different prices. In the future, this will be developed to the extreme by commercial interest groups. The biggest change in the block chain is to the supply chain. In the future, e-commerce will use block chain technology. There is no need to worry about fakes. Every product can be traced and traced. This will greatly enhance the competitiveness of e-commerce relative to physical business. If you look a little further, you can see from the cloud computing and big data above that user privacy is being violated more and more. Under the inducement of commercial interests, it is difficult for us ordinary consumers not to show their cards to these platforms. But the magic is one foot high and one foot high. Among these technologies, the block chain is one that can be used by ordinary consumers to protect their privacy. If our data is stored on the block chain in encrypted form, and any merchant can only use the data with our authorization, then the rights and interests of our ordinary consumers will be well protected[8].

There are many points that artificial intelligence can be combined with e-commerce. For example, artificial intelligence can be used to carry out more precise logistics and warehousing arrangements. If there are more IT elites in a certain area, then the warehousing in this area can be more IT electronic Products, which can significantly improve the buying experience in this area. Another example is the use of artificial intelligence technology to make more accurate portraits of consumers, and then the targeted display of different products will definitely increase the order rate, which can not only improve the user's shopping experience, but also greatly increase the platform's revenue[9]. The maturity and application of these technologies will significantly improve the efficiency of e-commerce in general. The combination of technology will improve the quality of services while reducing the number of service personnel [10].

$$\Phi(x, c_i) = \frac{1}{\left(x^2 + c_i^2\right)}, i = 1,2,\cdots,k \tag{1}$$

C is the center of the neuron, which is activated only when the given X is close to C, so the function has local induction. The output form of the neural network is:

$$F(x) = w_0 + \sum_{k=1}^{k} w_k \exp\left(\frac{x - c_k}{\sigma_k^2}\right) \tag{2}$$

Among them, K is the number of neurons in the hidden layer, and W is the bias.

2.3 E-commerce transaction mode and development strategy

E-commerce is divided into different e-commerce transaction modes according to different participants [11]. B2B e-commerce is the transaction of products, services and information between enterprises and enterprises through the Internet. Traditional companies generally adopt this B2B model, which can reduce inventory and procurement management costs and maximize corporate profits. B2C e-commerce refers to the e-commerce between enterprises and consumers. This business model is generally based on the online retail industry, using the Internet to carry out online sales activities, and realize mass consumption and provide services through the Internet[12]. The B2C transaction process has completely changed the traditional way of shopping. Consumers can safely buy what they need online without having to face-to-face. C2C e-commerce is a consumer-to-consumer transaction. The development strategy of e-commerce can be based on e-commerce, with the help of a macro analysis of the internal and external environment, examining the status quo, and fully grasping its own advantages and disadvantages. The development of e-commerce will involve many types of business activities, financial activities, etc.[13]. The domestic e-commerce strategy has generally gone through the stages of simple imitation strategy and Internet technology strategy. The development strategy theory mainly includes the strategic vision, and the emphasis is on what direction the company will develop into what kind of company in the future. The strategic goal emphasizes what kind of development state the company should achieve and should achieve in the future. The business strategy emphasizes which customers, which products and how to develop the company's future development. Functional strategy emphasizes what measures and strategies the company needs to take in marketing, research and development technology, etc. Because companies have various resources such as knowledge and technology, they have unique capabilities [14].

3 Experiments on the research and development of e-commerce strategic transformation under the background of the internet

3.1 Experimental background

The scale of the Internet industry continues to expand in China. Only 10 years after the Internet entered China, the number of Internet users has surpassed that of the United States. E-commerce relies on the rapid development of the Internet, and the Internet and e-commerce are also changing people's lives step by step. In the past two years, in order to expand their competitiveness, e-commerce companies have made ST one after another, but the effect is not too obvious. How to quickly realize the ST of e-commerce is a very realistic and very important issue before us, and it is also a very urgent issue. Therefore, this article conducts research on the development of e-commerce ST under the background of the Internet.

3.2 Experimental process steps

This article describes theories related to the ST of Internet technology and e-commerce. In order to better explore the ST of the e-commerce industry under the background of the Internet, it mainly borrows the literature research method and the questionnaire survey method to carry out comprehensive experiments. The experiment is mainly divided into two key contents. The first half of the experiment uses the literature survey method to compare and analyze the status quo of domestic mainstream e-commerce. In the second half of the experiment, a questionnaire survey method was used to analyze the effect of e-commerce ST in the context of the Internet. The questionnaire survey subjects were 100 senior employees of e-commerce companies, and the content of the survey was the effect of Internet technology on the ST of e-commerce.

4 Experimental analysis on the research and development of e-commerce strategic transformation under the background of the internet

4.1 The status quo and comparative analysis of mainstream e-commerce at home and abroad

The development of the Internet makes our life more convenient. In recent years, e-commerce has made our life more rich and convenient. Up to now, more and more online shopping platforms have emerged in endlessly, with batch after batch of online shopping apps, the scale of online shopping trading market is also expanding, and the transaction volume of e-commerce is also increasing. The current situation of transaction volume of several mainstream e-commerce platforms in 2020 is shown in the table below.

Table 1. The status quo of mainstream e-commerce at home and abroad.

	Operating model	2020 turnover
Alibaba	B2B	11450
Jingdong	B2C	3790
Pinduoduo	B2C	2420
Amazon	B2C	5750

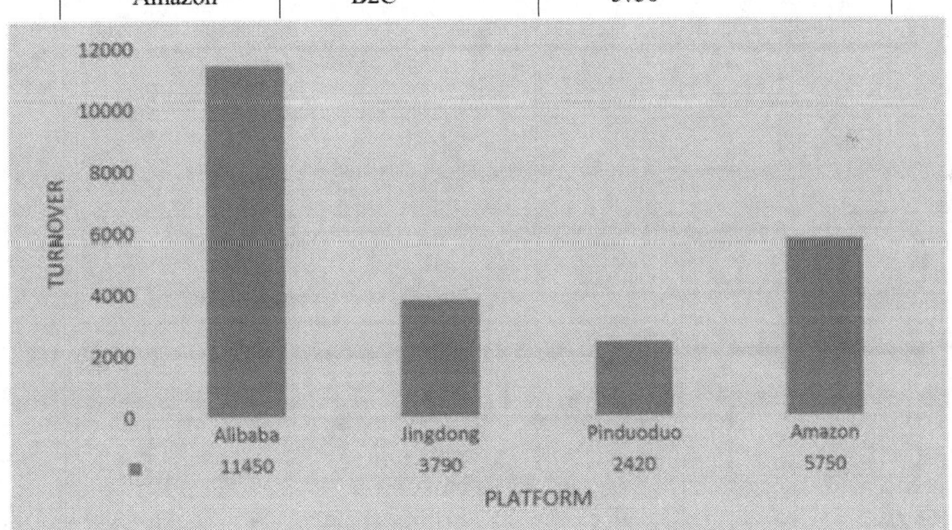

Fig.1. The status quo of mainstream e-commerce at home and abroad.

Alibaba, jd.com and pinduoduo are well-known e-commerce shopping platforms in China. They are known as the three giants of the e-commerce business community. As can be seen from table 1 and figure 1, as of December 31, 2020, Alibaba's annual turnover was USD 114.5 million, ranking first in the list of e-commerce transactions in 2020. In 2020, Jingdong's turnover was 3790 trillion yuan. The turnover of pinduoduo, a newly developed platform in recent years, is second only to JD in 2020, reaching 24.2 million yuan. The turnover of the three domestic e-commerce platforms reflects that the purchasing power of China's per capita consumption is very strong, and the development of the e-commerce industry is still optimistic. At the same time, foreign e-commerce platforms are not inferior. In 2020, Amazon ranked second with a turnover of

US $575 billion, second only to Alibaba enterprises. Among these four enterprises, only Alibaba is the business model between enterprises, which connects enterprises with each other. Jingdong, pinduoduo and Amazon adopt the business model between enterprises and consumers and strengthen the relationship between enterprises and consumers.

4.2 Analysis of the effect of e-commerce strategic transformation

Table 2. E-commerce strategy transformation effect.

	Number	Proportion
Excellent logistics	44	44%
Complete supply chain	21	21%
Innovative service concept	13	13%
Diversification strategy	22	22%

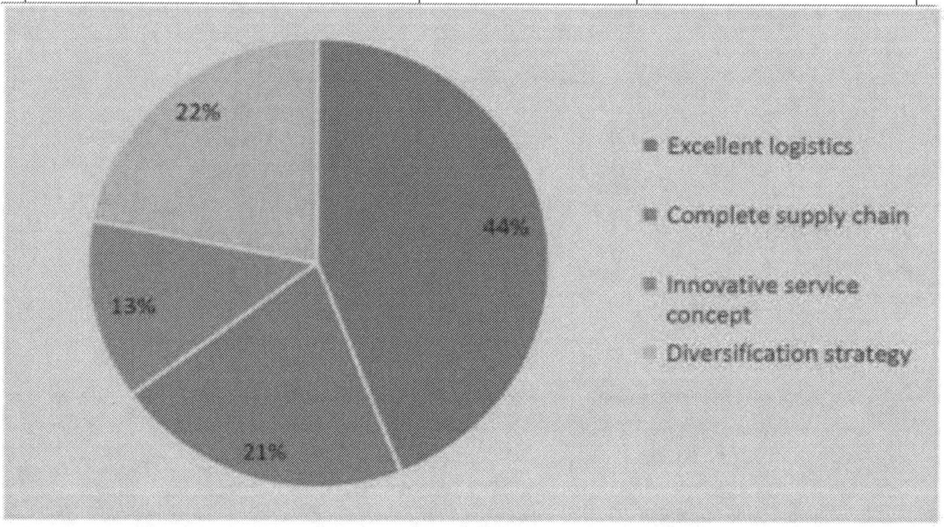

Fig.2. E-commerce strategy transformation effect.

It can be seen from the Table 2 and Figure 2 that 44% of people believe that the rapid development of the Internet provides an excellent logistics system for the e-commerce industry. The new logistics system uses information extraction, data mining and other technologies to ensure that orders are sorted automatically and accurately. 21% of the people also pointed out that e-commerce companies under the Internet background have a complete supply chain system, which is conducive to achieving their own differentiation strategy. For example, JD Business enterprise differentiation. 13% of people also feel that the development of Internet technology and information technology will help e-commerce companies provide users with special services and achieve their own differentiation. The remaining 22% also pointed out that the rapid development of the Internet will help e-commerce companies implement diversification strategies. The company has developed from initially focusing on the sales of one product to a comprehensive e-commerce company covering various categories. The relevant diversification strategy has given E-commerce companies have brought opportunities to transform the strategic matching of different product value chain activities into competitive advantages.

5 Conclusion

In today's world, e-commerce is always influencing people's life choices and changing the way people work. The mobilization and social networking of online shopping provide new business opportunities for e-commerce companies. Many e-commerce companies are only a few years away from establishment to exit. The reason is that some e-commerce companies ignore their own ST, do not have clear goals, and are gradually eliminated. For e-commerce companies, continuous transformation in line with market trends is the only way to go. Therefore, this article focuses on the ST of e-commerce in the context of the Internet, hoping to provide suggestions for promoting the diversification and differentiation of domestic e-commerce companies.

References

1. Huang P. Research on the path of transitional development of liquor industry based on the background of 'Internet +'[J]. Ecological Economy, 2017(02):172-180.
2. Jit S M. Clickstream analysis methods and systems[J]. Asian Journal of Management Sciences & Education, 2017, 1(3):57-65.
3. Liu Y. Research on the operating mechanism of China's coal market[J]. Agro Food Industry Hi Tech, 2017, 28(1):3075-3077.[4] He K. Development Strategy of "New Medicine" Under the Background of Healthy China[J]. Strategic Study of Chinese Academy of Engineering, 2019, 21(2):98-102.
4. Maintenance and development of agro-forestry-pastoral system from the perspective of biodiversity—Taking Zhagana in Diebu County as an example[J]. IOP Conference Series: Earth and Environmental Science, 2021, 836(1):012024 (8pp).
5. Maksimenko I, ashko T, drestova-Zakharenkova S. Digital transformation and its challenges to the strategic management system[J]. SHS Web of Conferences, 2021, 106(2):01035.
6. Druzhynina V, Likhonosova G. STRATEGIC IMPERATIVES ENSURING POPULATION WELFARE UNDER TRANSFORMATION EXCLUSION CONDITIONS[J]. Baltic Journal of Economic Studies, 2017, 3(5):135-145.
7. The Impact of Energy Internet Enterprise Construction on Power Grid Business Organization Mechanism[J]. IOP Conference Series: Earth and Environmental Science, 2021, 829(1):012002 (4pp).
8. Rajamanthri S. Current trends of HRM towards Effectiveness of the organisation[J]. Archives of Business Research, 2021, 9(5):1-10.
9. Ocakci E, Niemann J, Luminosu C, et al. An Overview of the Production Capacity and Operational Optimization in the Framework of Industry 4.0[J]. MATEC Web of Conferences, 2021, 343(12):05009.
10. Grechkosiy I, Osychka O, Yasynska S. Modeling as a tool for strategic management of effective economic recovery of a modern enterprise[J]. Socio-Economic Research Bulletin, 2020(3-4(74-75)):74-87.
11. Mukha S V. Modern Ukrainian Corporate Culture and Transformation of Business Processes[J]. Business Inform, 2019, 5(496):204-209.
12. Zsigmond T, Renáta Machová, Annamária Zsigmondová . Strategic Management from the Perspective of SMEs Operating in Service Sector[J]. Quality Innovation Prosperity, 2021, 25(2):37-53.

13. Chen P, Tang J W. Research on the Development of Micro-Tourism Market in Changsha City Under the Background of the Internet Era[J]. Asian Journal of Social Science Studies, 2019, 4(4):6.

Artificial Intelligence Technology to Explore the Future Development of Traditional Handmade Leather Goods

Zirui Qi

Beijing Institute of Fashion Technology, Beijing, China, 100029

Abstract. In recent years, artificial intelligence technology has developed faster and faster, and many industries need to rely on artificial intelligence technology for transformation and upgrading. Therefore, research on artificial intelligence technology is of great significance to the development of traditional handicraft industry. The purpose of this article is to study the future development and exploration of artificial intelligence technology in traditional handmade leather goods, and to specifically propose the use of intelligent new technologies to enhance personal experience, use new Internet media to innovate traditional handmade leather goods display, and expand composite types. The number of talents and other suggestions to promote the development of traditional handmade leather goods. Combining the problems of the personalized handmade leather goods style system, the decision support system mode is adopted to effectively combine the traditional handmade leather goods quantification, feature extraction and traditional handmade leather goods recommendation. In-depth research on user preference model and handmade leather goods recommendation algorithm, and established a model of personalized traditional handmade leather goods style system. Experimental results show that 9 out of 10 users are satisfied with the handmade leather goods recommended by the system after 5 interactions.

Keywords: Artificial Intelligence, Traditional Handmade Leather Goods, Future Development, Exploration and Research

1 Introduction

The cross-border integration of design and technology has played an important role in the renewal of traditional art [1]. From the perspective of exploring new process forms and compatibility with a wider range of product requirements, the integration of intelligent applications in the design and manufacturing process of traditional process products can increase productivity and productivity, and increase the interaction between products and users [2-3]. The manufacturing and dissemination methods of traditional handmade leather goods promote the transformation and development of the traditional handmade leather goods industry. From the perspective of art products as a means of traditional media communication, the creation of art culture means new designs based on 3D printing, intelligent integration, virtual reality, augmented reality, and larger intelligent application technologies. According to this idea, handicrafts can become elements of a new culture, defining culture in a new interactive way [4-5].

Nowadays, with the acceleration of globalization, traditional art has embarked on the road of commercialization, instrumentalization, and resourceization [6]. Some scholars argue that in a hyper-mechanized world, as a cultural heritage of mankind, traditional art should develop in the direction of restoring the humanity of mankind as a whole and ensuring cultural diversity. In addition, traditional art education should go beyond the purpose of preservation and inheritance, and expand its purpose with the source of restoration, practice, reflection, expansion and creativity of the human society as a whole. Whether traditional art will once again embark on the path of instrumentalization, commodification, and resourceization, or a channel for the restoration of human nature, this is the subject of human society as a whole, and the traditional art world in the new era also needs a moment of reflection[7]. Martin D conducted an in-depth analysis of the characteristics of traditional stippling and linked it to the common practice of non-photorealistic stippling and the capabilities and limitations of existing printing and display technologies. Based on the characteristics of traditional stippling (depending on the type of pen and paper used), we will focus on the perceptual research of digital stippling, which will provide further information for our discussion and work to understand the requirements of manual stippling reproduction. In order for artists and illustrators to faithfully replicate the stippling process in the digital realm, we extracted some guiding principles from the research results. For example, the characteristics of real points must be replicated because they are perceptible and must be based on the output device. Characteristic adjustment results [8]. The use of artificial intelligence technology can enhance the connection between people and products and between people, create sustainable culture and hot results, and form a relatively stable cultural communication cycle.

The research topic of this article mainly focuses on the combination of traditional handmade leather goods and smart material technology. The research object includes two levels: the first level, the research object is the product design of traditional handmade leather goods. On the basis of preserving the cultural characteristics of traditional handmade leather goods, improving functional practicability and interactive interest of traditional handmade leather goods, and meeting the needs of modern life, explore how to rationally combine smart materials with traditional handmade leather goods. At the second level, the research object is the intangible cultural dissemination of traditional handmade leather goods. Explore how to use traditional handmade leather goods as a carrier to provide a communication channel between media for the intangible culture based on smart materials. At the same time, connect products and networks, transition from "human-computer interaction" to "human-computer interaction", create a relatively stable cultural circle, and shape sustainable cultural exchanges.

2 Research on the future development and exploration of artificial intelligence technology in traditional handmade leather goods

2.1 CAD system for handmade leather goods

So far, the CAD simulation leather product system is a design assistance system, and its design direction strategy is to use an interactive system to provide designers with flexible and effective design tools [9-10]. Cognitive processes and technologies in the field of computer technology, such as computer science, engineering, related lighting and emission technology, fitness programs, etc., have not been successfully used in leather CAD manual systems. With the advancement of computer functions and the continuous improvement of CAD technology for handmade leather goods, on the basis of auxiliary design, combined with machine learning, intelligent technology and other technologies and technologies, it will improve the understanding of CAD system simulation of handmade leather goods [11]. Levels inspire inspiration Design inspiration, stimulate creativity and imagination. The use of intelligent technology can reduce the operating difficulty of the system and improve the efficiency of the system. Therefore, this technology is also one of the development guidelines for manual CAD leather products [12].

2.2 Promote the development of artificial intelligence traditional handmade leather goods industry

(1) Use smart new technologies to enhance personal experience

Use new technology to attract new people, use new artists to spread, attract new audiences. By incorporating elements of the traditional handmade leather goods industry into online games, and experiencing and watching traditional leather products through AR technology, these methods are in line with the way young people are exposed to new things and spread traditional culture through these exciting new technologies. This can not only promote the inheritance and development of the handmade leather goods industry, but also increase the cultural self-confidence of the younger generation.

(2) Expand the number of compound talents

The rapid growth of traditional leather goods requires the training of talents so that they can learn artificial intelligence while learning traditional leather goods. The expansion of traditional handicraft talents can mainly start from the following aspects: First, establish an effective government regulatory framework, provide certain support in terms of wages, wages, promotion and publicity, and encourage college. Students use the acquired knowledge and skills to enhance the value of traditional crafts and the ability to use artificial intelligence; secondly, they are connected with universities and give full attention to the effective talent courses of the Talent Academy, thus forming a cultural tradition. In the college simulation art, students who love traditional handmade leather goods are recruited to learn design and become the best successor of traditional handmade leather product simulation.

(3) Innovate the display of traditional handmade leather goods with the help of new online media

Based on the existing Internet, we eagerly use the new media "micro-three-terminal" as the medium to innovate the way of displaying handmade leather products. Using these new media to promote handmade leather goods can not only highlight the cultural significance of the design more clearly, but also more accurately highlight the many characteristics of the technology, culture, art and products of customized leather goods.

3 Investigation and research of artificial intelligence technology on the future development of traditional handmade leather goods

3.1 Traditional handmade leather goods design system

The development tool uses Microsoft Corporation Microsoft.NET. It is a new generation of Internet computer model launched by Microsoft. XML is an extensible markup language that has become a standard for data exchange. Web Services is a new idea proposed by Microsoft. It uses SOAP (Simple Object Access Protocol) and loosely coupled RPC mechanism as implementation methods, allowing different devices on the Internet to interact. Therefore, it solves the problem that the previous media such as DCOM, CORBA, RMI can not well solve the long-distance communication under the Internet environment. In the future, the software and services provided by many websites will become part of Web Services, which greatly saves development costs and improves the software reuse rate within the Internet.

3.2 User preference model and traditional handmade leather goods recommendation algorithm

The definition of a user preference model is a digital description of a specific preference from the submission of the user to the exit, including updating the user preference model as the user's preference changes.

This document represents the user preference model as two P:P (I, F) blocks. Among them, I is an array and F is a column vector. Each vector of series I is a chromosome in the result of genetic algorithm, and F represents the applicability of the corresponding chromosome. As shown in formula 1:

$$P = \begin{pmatrix} 0101 & 0101 & 0101 & 0101 \\ 0010 & 0010 & 0101 & 0101 \\ 1001 & 0101 & 1101 & 0011 \\ 1101 & 0110 & 0010 & 1110 \end{pmatrix} \begin{pmatrix} f1 \\ f2 \\ f3 \\ f4 \end{pmatrix} \quad (1)$$

Take the first 4 chromosomes to form the preference table I, and F is composed of the fitness function value of each corresponding chromosome as the weight of each sequence vector in the preference table. The expression of the preference model is shown in formula 2:

$$P = \begin{pmatrix} a_{11} & a_{12} & a_{13} & a_{14} \\ a_{21} & a_{22} & a_{23} & a_{24} \\ a_{31} & a_{32} & a_{33} & a_{34} \\ a_{41} & a_{42} & a_{43} & a_{44} \end{pmatrix} \begin{pmatrix} f1 \\ f2 \\ f3 \\ f4 \end{pmatrix} a_{ij} (i = 1,2,3,4; j = 1,2,3,4) \quad (2)$$

Get the similarity q (that is, the similarity of the two chromosomes, the similarity q is 0 by default). If xi = aij, then q = q + the influence of the accessory on the style multiplied by fi (i = l, 2,3,4; j = l, 2,3,4), the comparison of handmade leather goods will not be made in the future Components. The dress code is compared with the 4 lines of the preference model, and finally compared with the similarity p. Find the following handmade leather goods of this style and compare the following handmade leather goods. Until all the handmade leather goods of this style are compared, they are classified according to the similarity p of the handmade leather goods, and the one with the greater similarity ranks first. The system provides users with top-level handmade leather goods.

4 Analysis and research of artificial intelligence technology on the future development of traditional handmade leather goods

4.1 Digital style design of traditional handmade leather goods

A successful handmade leather goods style always has obvious style characteristics. From a specific design to a series of styles, the unity of style must be ensured. For handmade leather goods designers, one of the prerequisites for creating their own brand is to ensure that their design styles are basically the same, and the styles of handmade leather goods will also be carefully adjusted with the changes in popular elements of the times. Therefore, every element provided by the system can have style characteristics, and the system provides users with a style with typical style characteristics as a design reference, and provides certain guidance for users to select components, thereby inspiring users' design inspiration. The design work completed by the user using the design software should also be saved, so that the user can specify the style type according to his own style definition. After the user selects a specific style category and style type, the system first provides a variety of standard styles with style characteristics, and the user can replace components and some settings according to the standard style for quick and convenient smart style design. The style design process is shown in Figure 1.

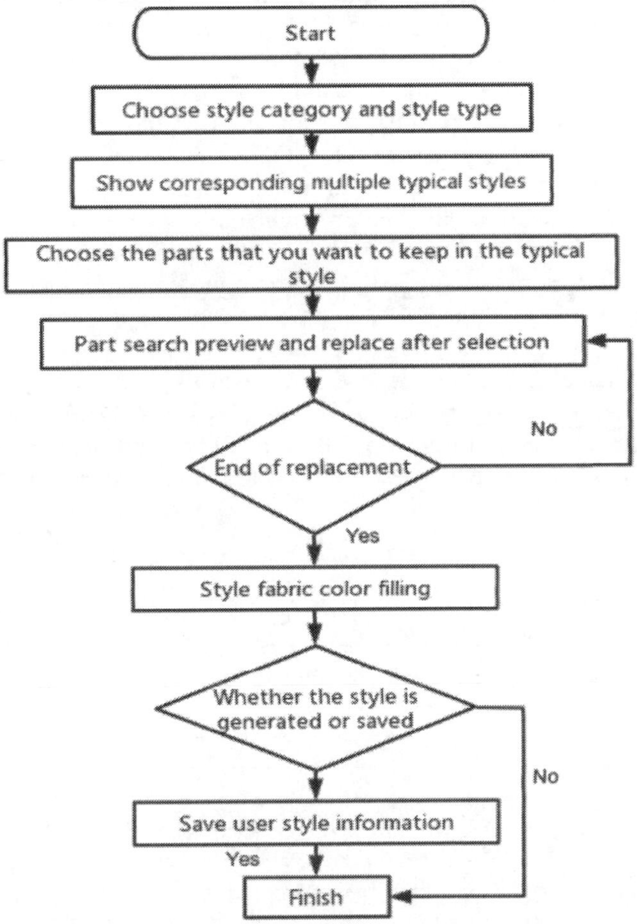

Fig.1. Style design flow chart.

The digital style design is based on a large number of component materials. At the same time, the system should provide software users with a reference for stylized design, as well as a wealth of standard style materials. Using the user style library provides users with a window to create their own stylized and defined designs. The popular season and popular location of handmade leather goods are also an indispensable basis for user design. Therefore, the food material library must frequently update the modern popular style, so that the designed works can have popular characteristics, closely integrate with the times, and be accepted by the society, and the works can be vigorous. Therefore, the style library of the system database should include features such as style type, popular time, and popular location.

4.2 Recommendation of handmade leather goods system

The system starts from the user interaction module. After the user logs in to the system, he first selects the style of handmade leather goods he needs, and then the system will present the handmade leather goods of that style in the handmade leather goods database to the user in a random order. Users start to score these handmade leather goods, and the number of handmade leather goods to be scored is up to the user. They don't want to continue scoring and just click to confirm. For some handmade leather goods that users dislike very much, it is not necessary to score them, and the information of the handmade leather goods will not be recorded. The user scored 50 pieces of handmade leather goods, and the system recommended handmade leather goods for the user based on the preference model. From all the styles of handmade leather goods selected by the user in the database, 25 pieces of handmade leather goods were recommended for the user and scored by the user, and the average score was 67.2. The user is not satisfied with the handmade leather goods recommended by the system and requires an update of the preference model. The system also requires the user to score each piece of handmade leather goods, in order to count the user's satisfaction, and the user can also view all the attributes of the handmade leather goods. If the user chooses to be satisfied, the current preference model is saved in the user database, and the whole recommendation process ends; if the user chooses not to be satisfied, the system will update the preference model and interact again. The system recommends handmade leather goods for him again. The 25 pieces of clothing users recommended this time scored an average of 89.2 points, and the users were basically satisfied. The following tests are performed on 10 users separately, and the user is first interacted with a single interaction without updating the preference model. The results are shown in Table 1:

Table 1. Results of a single interaction experiment.

User	The system recommends the average score of handmade leather goods	Number of handmade leather goods selected during interaction
1	86.4	32
2	87.4	36
3	92.8	55
4	90.4	48
5	98.4	60
6	83.7	30
7	66.8	25
8	88.1	35
9	94.6	57
10	80.3	28

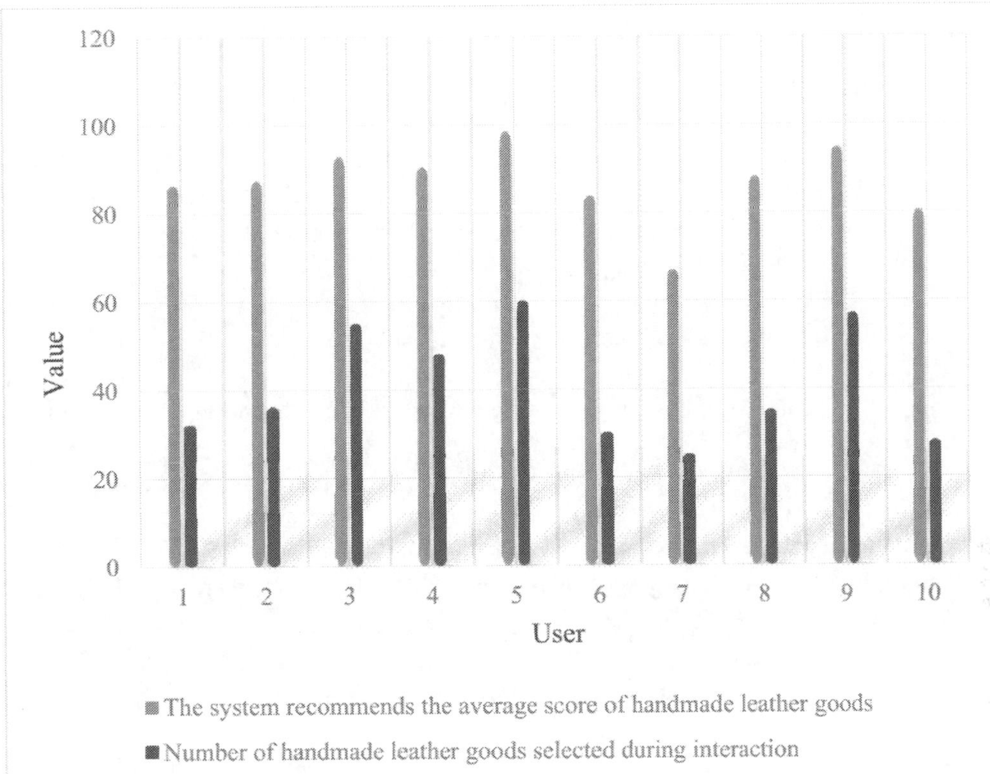

Fig.2. Results of a single interaction experiment.

In general, the more handmade leather goods a user chooses when interacting with the system, the final score is relatively high, indicating that the more user information the system obtains, the larger the initial population of the genetic algorithm, and the closer the result is to the user's needs, as shown in Figure 2 show. In the update test of the preference model, each user is set to interact with the system at most 5 times. If the user is not satisfied at this time, it is considered that the user's preference acquisition has failed. The recommended handmade leather goods are satisfactory, and one user is still dissatisfied after 5 interactions.

5 Conclusions

In the era of smart buildings, technologies such as artificial intelligence, Internet of Things, big data, and cloud computing are the core. While exploring and pursuing "technical advantages", it is also necessary to return to the human subjectivity represented by "technical spirit". The research purpose of this article is to start with the design of handmade leather goods, through the understanding of real user needs, to make logical innovations and upgrades to the life of handmade leather goods, integrate into life, and make a design close to "life". "Human-centered" life, summed up a logical and applicable structure, and combined with the structure of handmade leather goods. Make the design and implementation of leather goods more rational user design. This article takes authentic traditional handmade leather goods as an example. Basically, using artificial intelligence technology as the means and method of implementation can not only solve the needs of consumers, but also arouse people's attention to the traditional art of handmade leather goods.

References

1. Abramovitch I. Rise and Shinola The luxury brand's first hotel opens in the Motor City.[J]. Elle Decor, 2018, 29(8):114-114.
2. Moore P V. Artificial Intelligence: What Everyone Needs to Know[J]. Organization Studies, 2019, 40(3):466-470.
3. Chun H S, EH Kim. Traditional Korean Hand and Nail Beautification Culture[J]. Journal of Korean Traditional Costume, 2019, 22(1):29-45.
4. Chun H S, Kim E H. Hand and Nail Beautification Culture in the Wei, Jin, Northern and Southern Dynasties and Tang Dynasty[J]. Journal of Korean Traditional Costume, 2018, 21(4):43-57.
5. Connell W, Black M H. Artificial Intelligence and Legal Education[J]. The computer & internet lawyer, 2019, 36(5):14-18.
6. Borges A, Laurindo F, MM Spínola, et al. The strategic use of artificial intelligence in the digital era: Systematic literature review and future research directions[J]. International Journal of Information Management, 2020, 57(17):102225.
7. Komorowski M. Artificial intelligence in intensive care: are we there yet?[J]. Intensive Care Medicine, 2019, 45(9):1298-1300.
8. Martin D, Arroyo G, Sol V D, et al. Analysis of drawing characteristics for reproducing traditional hand-made stippling[J]. Computers & Graphics, 2019, 80(MAY):1-16.
9. Jfhg A, Hsh B. Tools from the World of Artificial Intelligence in Interventional Oncology: Be Careful What You Wish For - ScienceDirect[J]. Journal of Vascular and Interventional Radiology, 2019, 30(3):339-341.
10. Bin Y U, Kumbier K. Artificial intelligence and statistics[J]. Frontiers of Information Technology & Electronic Engineering, 2018, 19(01):6-9.
11. Burroughs D. The future of intelligence artificial[J]. International railway journal, 2019, 59(9):46-48,50-51.
12. Peng M, Quek T, Mao G, et al. Artificial-Intelligence-Driven Fog Radio Access Networks: Recent Advances and Future Trends[J]. IEEE Wireless Communications, 2020, 27(2):12-13.

Research on interactive innovation of computer-aided music teaching

Jing Wang, *Xiaolong* Deng

Xingtai University, Hebei, Xingtai, 054000,China

Abstract. Music education is an important subject that deepens five education and promotes students' art accomplishment and humanity cultivation in an all-round way, which is of great practical value to improve students' comprehensive quality. In the context of wisdom education, it will provide strong support for students to participate in music teaching deeply by giving full play to the advantages of computer-aided instruction and building an interactive platform for music teaching. By studying the value of interactive innovation in computer-aided music teaching, this paper puts forward the interactive innovation strategy of computer-aided music teaching.
Keywords: Computer; auxiliary music teaching; interactive mechanism; innovative strategy.

Music Education as a basic discipline, it is a basic quality of education. It reflects people's thoughts, emotions and social life by teaching music, an aural art, so that students develop strong interests and hobbies in music, having the ability of feeling, expressing, appreciating and creating the beauty of music, teaching the students how to know and feel life and create life in a certain sense, in order to achieve the goal of training students to develop in an all-round way.

1 The interactive innovation direction of Computer Aided Music teaching

The correct use of computer-aided instruction in all aspects of music teaching can directly appreciate the singing of musical works, performing posture and other forms of content, and the multimedia sound effects enable students to obtain a clearer musical image, the students have a strong emotional resonance. Thus, can deeply understand the music, the natural performance music, enables the student to obtain the emotion and the psychological satisfaction, the esthetic ability obtains the sublimation.

By analyzing human musical intelligence with big data, it simulates the information process of human visual, auditory, tactile, thinking and reasoning, and constructs its own neural network and algorithm generation, finally, it is applied to human perception music, cognitive music, research music, create music, and innovate the new music teaching mode of "Man-machine interaction". In the music lesson, the artificial intelligence system can simulate the scene of each big concert on the spot, real-time, enhances the student's esthetic appeal. Artificial intelligent electronic musical instruments have advantages that traditional musical instruments do not have. Electronic musical instruments have the ability to store a large number of musical instruments' timbres. On the basis of this, they can also arrange musical instruments' timbres, we all know that a good piano or violin costs tens of thousands of yuan to simulate the sound effect of a top-class musical instrument, but with simulated sound, we can solve this problem. In addition, with the virtual host, artificial intelligence accompaniment, 5g + VR/AR sharing platform and many other modern elements, to help students access the top music education experience.

The traditional composition theory has a modular teaching system, such as harmony, limbs and music paragraph structure, etc.. Musical artificial intelligence can not only simulate its teaching system, but also possess powerful, efficient and independent composing and calculating ability. The first AI music intelligence Composer, Orb Composer, introduces six basic music modules: Wind, strings, piano, electronics, Pop-Rock and Ambient to create music in a specific style in real time. Musical artificial intelligence can make people who know a little bit of music realize their dream of composing in an instant, but for people who have some ability to compose music, software can trigger their creative inspiration, the work will be automatically generated to do personalized, professional and meticulous modification.

2 The interactive innovative value of computer-aided music teaching

Music teaching is one of the important ways to implement aesthetic education. Therefore, we must pay attention to the characteristics of music in teaching and cultivate students' interest in music. Especially for the first-year primary school students, we should pay attention to the age characteristics of children. Teachers should create conditions in many ways, use modern educational technology means, colorful multimedia courseware let students play in secondary school, Music Middle School, Sports Middle School, learning Chinese music. Reasonable use of computer-assisted teaching, with a variety of teaching methods, to better develop students feeling, understanding, performance of music, from which the art of the edification, behavior of the United States. At the same time, through the collection of industry big data, but also to achieve personalized music education. The teaching platform in the distance music teaching environment can automatically collect the evaluation data of each player, form a personal learning growth data file, clearly understand the situation of piano teaching and scientific research, and through digital quantitative analysis, help the department to work out the best learning path. At the same time gather all the students of the school practice data, mining the most potential talent of piano performance.

The choice of network link is very important in the construction of distance music teaching environment, which directly determines the quality and effect of the course. In the construction of the music distance teaching environment, the biggest difficulty they encounter is the testing of the network link. The construction of the whole environment can be simply divided into audio-video interconnection and piano interconnection, in theory, two pianos could work as long as their links were clear over the internet. However, in the actual test, it was found that the common network link from the operator's computer room through the school's central computer room load balancing equipment and a series of firewalls, behavior management, switches and other equipment, QQ, wechat, web pages, audio and video can be linked to dialogue.

It has been proved by practice that "listening, thinking, singing, doing and editing" can be carried out in music teaching by using modern educational technology and integrating computer with music teaching, it is an effective teaching method in music teaching for imparting and training music knowledge and skills. By using this teaching method, students can acquire the basic knowledge and skills of music through training in different forms and contents, obtain the preliminary music feeling ability, the connoisseurship and the performance ability. The correct use of computer-aided instruction in all aspects of music teaching can directly appreciate the singing of musical works, performing posture and other forms of content, and the multimedia sound effects enable students to obtain a clearer musical image, the students have a strong emotional resonance. Thus, can deeply understand the music, the natural performance music, enables the student to obtain the emotion and the psychological satisfaction, the esthetic ability obtains the sublimation.

The network teaching mode not only enables students to directly experience the thinking mode and artistic creation of different regions and cultures, but also enables teachers and students not to leave school through the network exchange and cooperation among schools and organizations, we can experience the charm of foreign masters, and we can also interact with music masters and teams both at home and abroad through the Internet, so as to further enhance the inspiration of teachers and students. At the same time, the teaching under the network environment can integrate all kinds of scattered resources, gather and share all kinds of resources and intelligence, and play their respective strengths in the teaching interaction process. In the development of distance music teaching environment, the goal is to help teachers and students in the teaching and practice of playing repertoire, through the platform intelligence evaluation of players playing the piano pitch, rhythm, speed, strength, performance, difficulty and other dimensions of detection and analysis, through the functions of real-time feedback and accurate assessment, teachers and students can see the whole teaching process directly, and the assessment data are given to teachers and students when they play the piano, realize teaching students according to their aptitude and truly realize the closed loop of music teaching process.

3 The design of interactive innovation system of Computer Aided Music teaching

3.1 Structure of the musicianship training assistance system

The music skill computer-aided training system is composed of data acquisition, storage, analysis, display and decision-making. The input data of the system includes preset training content, data recorded by peripheral equipment and contrast data. The input is multi-channel and the output of the system is mainly image and sound. In addition, the system can be divided into three parts: the first part is a database-based training management system, including data entry and display, training progress management, the second part is an assistant system based on the analysis of practice data, which helps teachers to monitor the students' status and put forward suggestions on training content The third part is the system part, which is centered on the playback demonstration, to help teachers and students improve the training program. The training management system is the core of the whole system, which mainly corresponds to the database, the auxiliary analysis subsystem corresponds to the inner loop, and the playback suggestion subsystem corresponds to the self-correcting outer loop.

3.2 Functions and peripheral equipment of the CASM system

The system design needs to make the analysis to the music basic skill basic essential factor, take the vocal music training teaching as an example, from sings the physiological mechanism aspect to study. Vocal music is a singing skill, and vocal music training is a systematic training around this skill. Proper vocalization is coordinated by breathing, pronunciation, empathy, and language, but they are also interdependent, and no single factor can exist in vocalization alone. Our training process, whether from the breath of the hand or from the sympathetic position of the hand, should ultimately be the accumulation of the whole body in the best singing state. Therefore, CASM system in the application of vocal music training at the same time to take into account the sound and physiological aspects of many elements, which require different peripheral equipment and computer connection to collect. The sound module is mainly composed of recording equipment and its corresponding recording software, audio analysis software and integrated platform. The system can process the signals produced by the students' singing and playing in time. The sound is collected by recording equipment and software, and analyzed by spectrum analysis software, the music training elements, such as pitch, rhythm and resonance state, are extracted, and then the playback function is realized by recording equipment.

3.3 The main interface of the system

A large number of peripheral equipment in the system are connected to the computer, recording training data and displaying them in the main interface, the windows in the main interface have music window, reference wave window, training wave window, respiratory curve window, EMG curve window and record play button, etc. . These windows display different data for trainers to refer to. In the music library of the main interface, you can choose the songs you need to practice and the progress of teaching. Display the preset demo waveform in the reference waveform window, and the training waveform window displays the trainer's waveform. The breathing curve window shows the trainer's breathing condition during the training, and the myoelectric curve window shows the trainer's muscle tension condition. Various play buttons allow you to play demo recordings and trainer recordings. There are also buttons for rhythm analysis, pitch analysis, sentence break analysis and resonance analysis. These windows can be used to judge the accuracy of pitch, the accuracy of rhythm and intensity, the accuracy of pronunciation, the level of overtones in resonance and singing sound, the amount of breath and parameters of respiratory movement, the situation of larynx, the expression of musical expression, the trajectory of body movements in performance, etc. .

4 Sound module in 3.4 CASM system

The sound subsystem has three functions: real-time processing of sound signal and sequencer signal, comparing two kinds of signals and real-time feedback of signal difference. The system can also play the correct signal data preset in advance for learners' reference. The system can count out the critical errors, give the causes of the errors and improve methods, so as to achieve the goal of learning and training through testing. At the same time, considering the characteristics of testing and the requirements of learners, the system has real-time. This module involves the training content of sound, including rhythm and length, pitch, staccato, resonance and so on. Phrase graph comparison is a window that compares the waveforms of the referrer and the trainer by the sound waveforms, in which two sets of waveforms are displayed in real time. In the process of playing, students can see the waveform, on the waveform can directly see the volume size, the point of the lyrics, the rhythm of the graphics, the integrity of the phrase. By comparing the group information, the students themselves could find out the problem and then correct themselves. The sound of singing is a complex compound. The fundamental frequency of the human voice during singing is generally in the range of 100 Hz to 1500 Hz. In addition to the fundamental frequency, the song also contains a wealth of overtones and partials. Traditional teaching and training mainly rely on experience, using "functional hearing" to judge the sound of "chest, head, pharynx, oral cavity, high-position resonance" and other abstract concepts. After the spectral analysis of the sound, the pitch of the music and the intensity of various overtones, partials (timbre, resonance) in different frequency bands can be observed. This method is used to obtain the average spectrum, or "frequency spectrum", for the analysis of sound frequencies in a time domain.

5 Conclusion

Learning from the novel network cooperative teaching, we can fully learn from the characteristics of foreign well-known institutions of education, teaching features, use of transnational network links, automatic piano playing system, to achieve synchronous teaching in different places, making use of the network and automation equipment to create the simulated situation of field teaching can promote the emotion and infection of teachers and students, and help students to study effectively in the master class. Through the long-distance live broadcast, interaction, recording and playback of the course, while improving the teaching effect, it widens the boundary of region and time for music education, it has played a good demonstration effect in the integration of music education and science and technology.

References

1. Chen Feier. Talking about the application of computer music technology in music teaching [J]. Science and Technology Wind, 2017, (22): 26.
2. Fan Ling. The origin and evolution of interactive thinking in electronic music [J]. Journal of Nanjing University of the Arts (Music and Performance), 2017, (3): 107-114.
3. Yang Fan. On the Application of Computer Software in Harmony Teaching in Colleges and Universities [J]. Drama Home, 2017, (14): 191-192.
4. Niu Lin; Lin Yan. Thoughts on Using New Media to Spread Symphonic Music [J]. News Front, 2018, (20): 154-155.
5. Bao Wei. Application and Research of Computer Music Technology in Music Teaching in Colleges and Universities [J]. North Music, 2018, (8): 159-161.

Application of Information Technology in Chinese Language and Literature Teaching

Jinping Wang

Hebei Unversity, Baoding Hebei, China, 071002
Cangzhou Normol Unversity, Cangzhou Hebei, China, 061000

Abstract. The basic purpose of information technology is to promote, improve and support the teaching process. Language teaching technology is one of the products of communication and information technology in human science. It has a unique theoretical framework and principles. This technology is multidimensional and cross cutting. It is formed through the research on the theoretical and practical principles and basis of useful language level, computer language and knowledge of using the best technology. This paper discusses the problems existing in information technology teaching and information teaching environment, and finds that different types of teachers have different teaching methods.
Keywords: Information Technology, Chinese Language, Teaching Methods, Teachers

1 Introduction

With the continuous progress of science and technology, great changes have taken place in people's lifestyle and social model. Information technology (it) is booming at an alarming speed, which has a far-reaching impact on the education system of all countries in the world. Computer, Internet and other technologies are now being used by teachers and learners at an increasing speed. Institutions are exploring ways to make effective use of information technology for education.

With the continuous development of computer technology, many experts have studied language teaching. For example, Manalu H F study aims to understand how often students and lecturers use information technology in an English learning environment. A questionnaire was distributed among students and lecturers in the Department of English literature at the University of Bangladesh. The results show that lecturers and students have a positive attitude towards the use of information technology tools and think that using these tools in the teaching process is very useful [1]. Wang L, Xu Z constructed a wireless distance English teaching system based on mobile edge computing, and collected data from the database by means of case analysis and literature analysis. The teaching mode based on network teaching is established, and a large number of relevant documents are read and analyzed through literature research [2]. Sabiri K A discussed and analyzed the factors affecting pre service and in-service teachers and student teachers' views and confidence in ICT integration [3]. Although the research results of Chinese teaching are quite rich, there are still deficiencies in the application of information technology in Chinese teaching.

In order to study the application of information technology in Chinese teaching, this paper listened to the research on information technology and Chinese teaching, and established an automatic identification system of information points. The results show that information technology is conducive to Chinese teaching.

2 Method

2.1 Information technology

(1) Several problems in Information Technology Teaching

The implementation process of information technology curriculum is a process of symbiosis between real culture and virtual culture, between people and media and various information cultures, and a process of cooperative learning between teachers and students. The discipline characteristics of information technology determine that its educational environment is a harmonious field nourished by culture [4]. The existence of information technology classroom authority leads to the status differences between teachers and students in the process of teaching dialogue, which is mainly reflected in the authority of teachers. In the field of teaching, both teachers and students may be the transmitter of information [5]. However, in the current information technology classroom teaching in senior high school, most teachers still dominate the classroom, play an absolute authority and leadership role, and implement "localized" teaching, while teachers are in a strong position in discourse communication. Their application in information technology classroom teaching dialogue is in line with the teaching concept of the new curriculum and is affected by discourse hegemony, the authoritative image of teachers has also reached the extreme [6]. At present, in the process of information technology classroom teaching dialogue, most teachers are monologues. The communication between teachers and students is essentially one-way. There is almost no "real" interaction between the subjects of teaching dialogue, which plays a decisive restrictive role in the performance of students' dialogue [7].

(2) Information teaching environment

Information-based teaching environment is the sum of various situations and conditions in information-based classroom teaching [8]. It is a classroom teaching environment created by using modern educational theory and information technology. From the perspective of information construction and activity mode, China's school education information classroom teaching environment can be divided into multimedia classroom teaching environment, audio-visual communication environment and network teaching environment. Information-based teaching environment mainly refers to the multimedia classroom under the multimedia classroom teaching environment [9]. Its main components are multimedia podium, projector, projection screen and blackboard. Teachers can use computers and projectors to display teaching content on the projection screen, and use computers and audio-visual equipment to play video, audio and video, so that the whole class can observe and watch clearly and concretely at the same time. The application of task-based teaching method in the environment of information technology provides students with more opportunities to participate in class and practice [10]. Students use Chinese for meaningful communication in a real and natural language environment. Through communication, interaction, unity and cooperation, they complete various tasks in the target language, so as to master language knowledge and improve their comprehensive language application ability.

2.2 Chinese language and literature teaching

(1) Humanity of Chinese knowledge

Chinese is presented in the form of literary texts [11]. Every literary work is the author's spiritual product. Although it has no flesh and blood, it exists in a plump and round posture, but it has the same meaning of life. Then the humanity of literary knowledge in Chinese teaching is undoubtedly stronger [12]. We can see that the teaching contents and achievements of Chinese are language expression skills, but this skill permeates a strong humanistic spirit from bottom to top; the other part is the emotional thinking that we can't see. This is tacit knowledge, and the knowledge of Chinese language is the hidden language expression and application ability. Language expression and application ability. Besides reading and writing skills, listening and speaking skills are also very important. We should be obedient, get information from "listening", judge the language environment and language object, and choose the appropriate corpus when speaking, so as to form a unique language style.

(2) Chinese learning style

In the existing textbooks, we can find many written expressions, and there are still many characteristics in explaining the relevant grammar and vocabulary: the problems include difficult words, difficult understanding and difficult grammar. These problems pose great challenges to Chinese learning and improve students' understanding of Chinese learning. We can understand Chinese grammar from another angle, simplify and clarify Chinese grammar. For example, the explanation of professional grammar is replaced by relevant illustrations and relevant literature. In the process of classifying each category, several examples can be obtained to prove the previous knowledge, which not only makes learners understand each knowledge point more accurately, but also can better form learners' learning concept, better promote memory, and lay a solid foundation for long-term memory in the future. From another point of view, the previous teaching materials are more complex, and there should be more such expressions, so as to combine practice with teaching, better promote teaching and promote the learning of Chinese and culture.

(3) Introduction to teaching methods of different types of Teachers

Expert teachers also tend to use multimedia to present teaching content, but in terms of media function, they prefer to use multimedia to create situations, guide teaching content, stimulate students' positive attention and stimulate students' thinking in the form of problems. Expert teachers use information technology to implement the process of information teaching design and development, adopt reasonable teaching strategies, create real teaching situations, and pay attention to the cultivation of students' creative thinking. Through the situation created by teachers, students can find problems, put forward problems and solve problems. Teachers throughout the teaching process to explore the idea of ingenious teaching methods. Therefore, whether in the process of teaching design or classroom teaching, novice teachers should pay attention to strengthening the application of inquiry teaching, create a real teaching situation with the help of information-based teaching environment, help students find and solve problems from the situation, guide students to think deeply and deeply understand the teaching content.

2.3 Automatic Identification of Information points

After grouping multiple clusters, the system evaluates the clustering effect according to the following indicators to measure the clustering effect. These indicators are several verification indicators based on the distribution of class elements. These indicators include compaction verification index CMP, separation verification index SEP and comprehensive verification index OCQ (considering CMP and SEP). The degree of compaction (CMP) index is defined as formula (1):

$$\mathrm{WheCmp} = \frac{1}{c}\sum_{i=1}^{c}\frac{V(C_1)}{V(X)} \quad (1)$$

re C is the number of classes divided.

The keyword defined by Boolean model only appears or does not appear in the document, that is, the keyword weight has duality: $W_{i \cdot j} \in (0,1)$. The correlation between the document and Q is defined as follows (2) (q is a Boolean expression):

$$\mathrm{Sim}(d_j, q) = \begin{cases} 1, q \in d \\ 0, q \notin d_j \end{cases} \quad (2)$$

If $\mathrm{Sim}(d_j, q)$ indicates that the query statements Q are related to the document, otherwise they are not related.

When evaluating an information extraction system, the above two indexes must be considered to comprehensively evaluate the performance of the system. F indicator is an indicator established to evaluate comprehensive performance, and its calculation formula is as follows (3):

$$\mathrm{F-Measure} = \frac{(\beta^2+1)\mathrm{PrecisionRecall}}{\beta^2\mathrm{Precision}+\mathrm{Recall}} \quad (3)$$

Among them, β is a weight parameter that determines whether to focus on recall rate or accuracy rate during evaluation. It is preset by the system. The relative importance of accuracy and recall can be reflected by adjusting the β value.

3 Experience

3.1 Object extraction

Taking novice teachers as the research object, this paper uses the method of case study to observe the classroom teaching behavior of novice teachers in two different teaching environments. In the process of observation, in order to maintain the homogeneity of observation contents in the two teaching environments, a novice teacher is selected as the research object to avoid the impact of teachers' individual differences on teaching behavior; as far as students are concerned, the grades of the two classes remain the same, and the contents of the courses taught are the same. In this way, we should try our best to avoid the influence of other factors on Teachers' teaching behavior. Only classroom teaching environment can be used as the influencing factor of teachers' teaching behavior. After the experiment, the second questionnaire survey and on-site interview were conducted to the students in the experimental class. Finally, two students were tested for the second time. The purpose is to understand the changes of students' learning attitude and Chinese ability under the new teaching mode. Through the test results, we can understand the difference between task-based teaching method and traditional teaching method in the environment of information technology.

3.2 Experimental analysis

Firstly, it improves the automation requirements of the extraction system and the adaptability of the extraction rules. Due to the rapid growth of web pages, users constantly modify or add extraction rules according to the changing web pages, which is obviously inefficient. However, the current information extraction methods with high degree of automation have low accuracy and adaptability, poor practicability and good performance. Information extraction technology requires the participation of a large number of users, and the degree of automation is not high. How to solve the contradiction between them and realize the dynamic balance between them has become an urgent problem to be solved. At the beginning of the experiment, the students conducted a pre-test, including oral and written tests. Teachers collect relevant data to judge students' actual Chinese level. At the same time, the first questionnaire survey was conducted on the students of the two classes to understand the students' attitude towards Chinese learning, their interest and motivation in Chinese learning and the current situation of Chinese classroom teaching. At the same time, the students put forward their own views and suggestions on the reform of Chinese teaching. After the prediction test, the author uses different teaching modes in the two classes and carries out teaching work at the same time. Combined with Chinese teaching materials, the experimental class adopts the Chinese task-based teaching method under the environment of information technology to cultivate students' Chinese ability. The control class organized by the same teacher still practices the same Chinese materials according to the traditional teaching methods.

4 Discussion

4.1 Extraction of XML master data block

In order to reduce the complexity of extraction, the strategy first weights the data blocks, extracts the main data block, and then further extracts the data records of interest to users in the main data block. According to the data block weight algorithm, calculate the weight of the data block based on the bottom-up first common ancestor node as the div, as shown in Table 1.

Table 1. Div data block weights.

	body	TBODY	TD	DIV	A	Leaf node
Node weight	45	53	32	26	38	36
Block weight	47	48	36	83	73	53

It can be seen from the above that the node weight of the body is 45 and the data block weight is 47; the node weight of tbody is 53 and the data block weight is 48; the node weight of TD is 32 and the data block weight is 36; the node weight of is 26 and the data block weight is 83. The specific presentation results are shown in Figure 1.

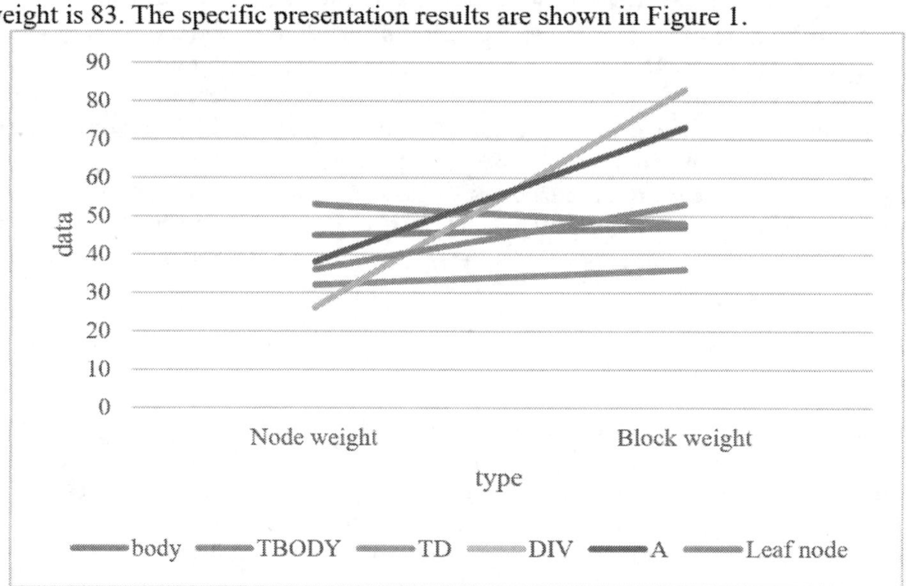

Fig. 1. Div data block weights.

As can be seen from the above figure, in the information display web page, the number of master data blocks is usually the number of blocks corresponding to the two largest weights, which is reflected in the main data block extraction of Div. The block corresponding to the two largest data block weights is the master data block (Note: the data block weight corresponding to the calculation label represents the data block weight under the label)

4.2 Recommendation strategy and principle of course synchronous reading materials

The course synchronous reading materials need to have enough vocabulary and text overlap. In the stage of basic education, expanding students' vocabulary is the key content of Chinese learning, and vocabulary is also the key factor affecting students' reading. Vocabulary reflects their reading ability. For primary and middle school students who are relatively lack of experience and social knowledge, especially primary and middle school students, it is necessary to strictly control the proportion of superwords in synchronous reading materials. In addition, high-level words should not be arbitrary, as shown in Table 2.

Table 2. Ebbinghaus memory curve.

Days	1	2	3	4	5	6
Percentage maintained	36.4	28.4	27.8	27.8	27.8	27.8

It can be seen from the above that forgetting begins immediately after learning, and the process of forgetting is not uniform. At first, the forgetting speed is very fast, and then gradually slow. After a long time, it is almost no longer forgotten. The specific results are shown in Figure 2.

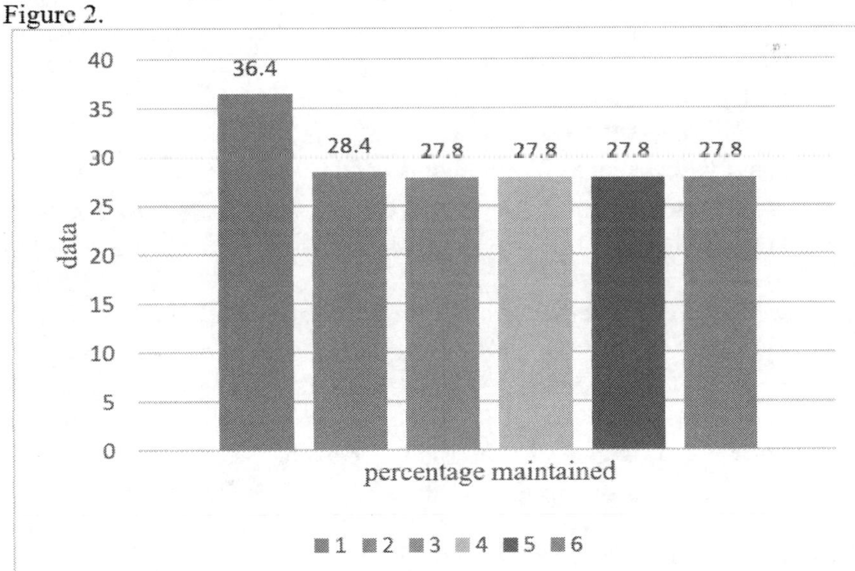

Fig. 2. Ebbinghaus memory curve.

The above results show that students' forgetting law should be considered when choosing synchronous reading materials. In our daily study, we often encounter the phenomenon of "rebirth of reading and writing". The so-called "reading and writing cycle" refers to the phenomenon that students' memory traces gradually disappear or are disturbed with the passage of time. In other words, the vocabulary of grade 6 can appear in the synchronous reading materials of grade 5, while the vocabulary of junior middle school should be strictly controlled.

5 Conclusion

With the continuous progress of society and the rapid development of information technology, the application of information technology has widely entered the field of education, which has a great impact on the concept, method and mode of education. Information technology has injected new vitality and vigor into education. The main purpose of this paper is to explore the latest development of information technology in language and literature research. The research results show that different types of teachers have different teaching methods, and the use of information technology undoubtedly provides great help for teachers in the teaching of Chinese language and literature.

Acknowledgments

Hebei University Humanities and social sciences research project "Research on the construction of Chinese language and character curriculum system in Local Normal Universities under the background of new liberal arts", Project No.: SD2021117.

References

1. Manalu H F. The Analysis of Information Technologies Usage for Academic Purposes in English Language Teaching and Learning at Universitas Bangka Belitung[J]. JOALL (Journal of Applied Linguistics and Literature), 2019, 4(2):177-184.
2. Wang L, Xu Z. An English Learning System Based on Mobile Edge Computing Constructs a Wireless Distance Teaching Environment[J]. Mobile Information Systems, 2021, 2021(1):1-9.
3. Sabiri K A. ICT in EFL Teaching and Learning: A Systematic Literature Review[J]. Contemporary Educational Technology, 2019, 11(2):177-195.
4. Mahzari M, Saad M, Ethleb H, et al. Learning Arabic as a Second Language: An Exploration of the Efficacy of Arabic Subtitles by Netflix Viewers[J]. The Iranian EFL Journal, 2021, 11(3):174-197.
5. Li W. Multimedia Teaching of College Musical Education Based on Deep Learning[J]. Mobile Information Systems, 2021, 2021(2):1-10.
6. Hashemi A, Si N K. The Barriers to the Use of ICT in English Language Teaching: A Systematic Literature Review[J]. International Journal of Information and Communication Technologies in Education, 2021, 3(1):77-88.
7. Jufri S, Apriyanti D. Acquiring English of Teaching Subject through CLIL in Teacher Training Bung Hatta University[J]. Journal Polingua Scientific Journal of Linguistics Literature and Education, 2019, 8(2):53-56.
8. Sushma B. Teaching Analytical Writing Through Editorials Of Newspapers[J]. International Journal of Scientific & Technology Research, 2020, 8(3):4711-4713.
9. Pieper I. The challenge of plurality. A comment on the occasion of the journal L1-Educational Studies in Language and Literature being 20[J]. L1-Educational Studies in Language and Literature, 2021, 21, Special Issue L1(SI Twenty years L1):1-5.
10. Yawiloeng R. Second Language Vocabulary Learning from Viewing Video in an EFL Classroom[J]. English Language Teaching, 2020, 13(7):76-83.

11. Yan Y, Sun X, Li B, et al. Construct a Teaching System Combining Image Linguistics and Multimedia Technology[J]. Wireless Communications and Mobile Computing, 2021, 2021(1):1-11.
12. Zhou Y. VR Technology in English Teaching from the Perspective of Knowledge Visualization[J]. IEEE Access, 2020, PP(99):1-1.

Research on Intrusion Detection Based on User Portrait Technology

Zhenduo Wang

Xi'an Siyuan University, Xi'an, Shaanxi, 710038, China

Abstract. Under the big data scene, the intrusion detection technology has some problems, such as weak data processing ability, difficulty of precise location and so on. The intrusion detection technology based on user portrait technology, on the basis of fully mining the practical value of big data technology, refines the detection granularity, improves the intelligent and accurate characteristics of intrusion detection. The experimental results show that the intrusion detection model based on user portrait can detect abnormal users quickly and accurately, and improve the effect of intrusion detection, and has good practical value.
Keywords: User profile technology, intrusion detection, data samples, user tags

At present, there are still many problems in the process of building user profile, such as user data collection, data processing, behavior data extraction and modeling, and finally user tag generation. This involves a lot of manual operation, at the same time artificial subjectivity will affect the objectivity and accuracy of the user's portrait.

1 An overview of intrusion detection technology based on user profile technology

Intrusion Detection System (IDS) is one of the important security measures in network security. It can provide real-time dynamic security protection for the system and take emergency response measures when threats are found. By establishing a user tagging system, the user's actions are extracted into short-text tags, the weights of each tag are set, and a series of tags are used to form a virtual user profile, and the anomaly indices among different users are calculated, complete the detection of abnormal users.

1.1 Overview of user profiling techniques

This paper introduces user portrait technology into security field, collects user operation data by itself, and extracts user behavior habit by user portrait. User portrait technology is based on the mathematical modeling of real-world users based on user data: In general, to build a user portrait requires data, based on users' social attributes, living habits and consumption behavior, etc. , the collection and accumulation of user data is the basis of user portrait; to have a clear business application scenario, user portrait and business application are inseparable, it is often necessary to profile and analyze specific users who meet the business requirements, and to have relevant user modeling algorithms to mine deep level information from existing user data that can touch user requirements, a tabbed user model that abstracts different data. The core work of user portrait technology is to label the user, and the label is a highly refined feature mark through the analysis of the user information, so as to classify and extract the user based on the label.

1.2 Overview of intrusion detection systems

Intrusion Detection technology can be divided into two categories: Abuse Detection (Misuse Detection) abuse Detection is to use the known intrusion methods and weaknesses of the system to identify illegal intrusion. The main disadvantage of this method is that all known intrusion patterns are embedded in the system, so any unknown intrusion can not be detected. But the detection efficiency of this method is high. Anomaly Detection is to detect whether the current user behavior deviates from the established normal behavior profile to identify whether there is illegal intrusion or ultra vires operation. The advantage of this method is that it does not need to know the system defect and has strong adaptability. But the possibility of false alarm is high. The user behavior in intrusion detection system is mainly in the form of data. According to the source of data, intrusion detection system can be divided into host-based and network-based two. The former comes from the operating system's audit data, and the latter from the packets that flow through the network. Since the behavior of users is data, the core of the problem is how to deal with the collected data correctly and efficiently, and draw conclusions from it.

2 Construction of intrusion detection system based on user profile technology

According to the rules of the rule base, intrusion detection detects the behavior of the current user, and adopts different countermeasures according to the results. In order to solve the problem that the user behavior will change periodically over time, the concept will drift and the false alarm rate will rise. In order to avoid the influence of historical transaction data to a certain extent, a sequential user behavior profile is proposed, and the diversity of future transactions is expanded, finally, the experiment is compared with other algorithms.

2.1 Evaluation index of Intrusion Detection System based on user profile technology

2.1.1 Detection rate: TP/(TP + FN)

It refers to the proportion of attack samples which are recognized as attack correctly, which is an important index reflecting the ability of IDS attack recognition

2.1.2 True negative rate: TN/(FP + TN)

The percentage of normal samples in the test set that are correctly identified as normal is an indicator of the accuracy of IDS in identifying normal samples

2.1.3 Accuracy: TP/(TP + FP)

Refers to all IDS identified as attack samples in the test set, the true ratio of the attack samples;

2.1.4 F-score:

A Comprehensive Evaluation Index of IDS detection rate and accuracy rate;

2.1.5 Classification accuracy: (TN + TP)/(TN + TP + FN + FP)

The ratio of the number of correctly classified samples in the test set to the total number of samples in the test set, which is an overall evaluation index reflecting the ability of IDS to distinguish between normal samples and attack samples, to a certain extent, can reflect the overall identification ability of IDS;

2.1.6 Omission rate: FN/(TP + FN)

The ratio of the number of attack samples misidentified as normal in the test set to the total number of attack samples in the test set is an index which can reflect the ability of IDS to identify attacks

2.1.7 False alarm rate: FP/(FP + TN), also known as false alarm rate

The ratio of the number of normal samples in the test set to the total number of normal samples in the test set is an indicator of IDS' ability to recognize normal samples.

2.2 Content of intrusion detection system based on user profile technology

Network-based IDS can not provide the service of intrusion detection separately. At first, IDS was chosen because of its low cost and quick response. The policy configuration of network-based IDS has several key points for multiple systems and can observe the network communication of multiple systems, so many hosts do not require installation when managing and loading software. The configuration environment with fewer monitoring points is the network environment with lower cost. Network-based IDS will detect all the packets passing through, and will find some malicious programs and software, also can not detect the corresponding type of attack. For the network-based IDS can detect some effective load content, can detect some specific attack tactics and instructions.

By checking the packet to see if it has payload capability, and can view the malware, it can not be detected by an attacker, but host-based IDS can not detect its behavior, unable to identify the appropriate attack message. The attacker will not easily transfer the evidence, in the process of real-time monitoring of the attack, the network-based IDS will use the network to communicate. There are many ways to capture evidence, including methods of attack, which can identify the identity and information of a hacker, as well as record basic information and other bad behavior, do not give any opportunity to cover up the traces of crime, to prevent the use of this information to host-based system for intrusion detection.

Network-based IDS can detect suspicious malicious operations at any time and any place, and can make fast notification and response. TCP is one of the network protocols, a TCP based denial of service attack on the network, can send an emergency signal through IDS, make TCP Reset. For host-based systems, real-time monitoring is incomplete, and only suspicious information can be monitored and recorded during operation, so that its attack capability can be identified and corresponding response can be made, the system may have been compromised, or the host-based IDS may have been destroyed and a lot of information may have been leaked. Network-based real-time monitoring, can be real-time notification, according to the specific circumstances of the notification to make rapid response to prevent information leakage, resulting in losses. Network-based IDS data adds so much valuable data that it's hard to tell which ones are bad intentions, even if the network's firewall can resist these attempts, the network-based IDS outside the firewall can also detect the intention of the attack behind the firewall. The detection of the above data plays an important role in evaluating the security performance of the system.

2.3 Application steps of user behavior profile in intrusion detection

Host-based intrusion detection system (ids) is not so fast compared with network intrusion, but it does have some advantages that the local area network (Lan) system can not match, better identification analysis, close attention to particular events, and low cost. The first step is to determine whether the attack is successful or not. The host-based IDS uses information that contains events that have occurred, and it can more accurately determine whether the attack is successful or not. The host-based IDS is a perfect supplement to the network IDS. The network part can send warning message as early as possible, and the host part can decide whether the attack is successful or not.

Monitor specific system activities, monitor user and file access activities based on the host IDS, access to files, changes to file permissions, creation of new executable files, access to licensed services, and so on. Host detection can detect the attack that network system can not detect. Host-based systems can detect attacks that can not be detected by network-based products. Suitable for encrypted and switched environments, host-based systems can be installed on each host, more suitable for switched and encrypted environments than network-based detection systems. Switching devices can manage many large networks by dividing them into many small networks to cover a sufficiently large network range and perspective.

2.4 Method verification and analysis of intrusion detection system based on user profile technology

A user portrait is a virtual representation of a person, based on a set of data models. Based on the mobile phone perception data, the user portrait was constructed from age, sex and personality. It uses sensors and event listeners in the phone to collect data from the slide-screen unlock scenario, basic information about the phone, application usage, and screen-state scenarios. The quality of training data in data mining directly affects the accuracy of extracted user features and derived rules. If the intrusion detection system includes the behavior of the intruder in the data used to build the model, the detection system will not be able to respond to the intrusion, resulting in under-reporting. The data used for training must not contain any intrusions and must be formatted in a form that the data mining algorithm can process.

The system's data source is network-based, using sniffers to capture the user's packets, and then using protocol analysis to discard the payload and keep only the Baotou portion, the data pre-processed by a specific method consists of 7 fields: time, source IP, source port, Destination IP, destination port, connection ID, connection status. Because TCP connection building involves three handshakes, all the training data collected will include some unsuccessful connections that will negatively impact the subsequent data mining process, therefore, only those data that reflect the normal state of the network should be retained. This is not the case with UDP, where each UDP packet is treated as a single connection.

Using data mining technology in intrusion detection system, we can extract user's behavior characteristics and summarize the rules of intrusion behavior by analyzing historical data, so we can set up a complete rule base for intrusion detection. Data collection network-based detection system data from the network, available tools such as TCPDUMP. By writing a Python program to collect user behavior data to form a user profile, the detection model is established. The model extracts the features of the user behavior, uses the machine learning method to learn the normal user behavior, and uses the mahalanobis distance and isolated forest algorithm to determine whether the tested behavior is abnormal or not.

3 Conclusion

The existing user search portrait technology has the problem of large feature dimension and sparse matrix. User portrait technology for massive user behavior data still needs more complex and mature algorithm research to promote. User portrait technology is a combination of multi-disciplines, which needs knowledge fusion in knowledge mapping, natural language processing, machine learning and data mining. Using data mining technology in intrusion detection system, we can extract user's behavior characteristics and summarize the rules of intrusion behavior by analyzing historical data, therefore, a more complete rule base is established for intrusion detection.

References

1. Li Junzheng, Huang Hai, Huang Ruiyang. et al. Research on user search portrait technology based on chi-square test and SVM [J]. Electronic Design Engineering, 2017, (24).6-10.
2. Wang Le, Ni Weijian, Lin Zedong, et al. User portrait method of online behavior logs based on model stacking [J]. Journal of Shandong University of Science and Technology (Natural Science Edition), 2018, (5).70-78.
3. Xu En, Yu Zhiwen, Du He, et al. User portrait system based on mobile perception data [J]. Journal of Zhengzhou University (Science Edition), 2019, (4).30-36.
4. Guo Na, Wei Rongkai, Shen Yanping. Abnormal Feature Extraction in Big Data Environment Based on User Profile [J]. Computer Simulation, 2020,(8):332-336.
5. Chen Si, Wu Qiuxin, Zhang Mingkun, et al. Intrusion detection technology of intelligent industrial control system based on edge-cloud collaboration [J]. Computer Application and Software, 2020, (11): 280-285+333.

Research on the Teaching Mode of Ideological and Political Education with the Integration of Information Technology and Teaching

Dong Wang, *Xia* Zhao, *Hongwei* Wang

Ordos Vocational College of Eco-Environment, Inner Mongolia, Ordos, 017000

Abstract. In the context of wisdom education, actively promote the application of modern information technology and ideological and political education and teaching, forming a new ideological and political education ecology, by giving full play to the advantages of the application of new media and new technology, it makes ideological and political teaching more vivid and flexible, and fully promotes the attraction and interest of ideological and political teaching. Based on the analysis of the logical basis of the teaching integration of ideological and political courses, this paper puts forward the predicament of the teaching integration of information technology and teaching and puts forward some suggestions.
Keywords: Information technology; teaching integration; ideological and political course; teaching mode

Taking the application of information technology in ideological and political course teaching as a starting point, combining with the age and psychological characteristics of college students, this paper analyzes the difficulties and opportunities faced by ideological and political course teaching in colleges and universities, finding problems from the present situation of the application of information technology in ideological and political courses in colleges and universities, pushing forward the reform of ideological and political courses in a targeted way, stimulating the inner impetus of ideological and political courses, and promoting the sense of gain of college students and the sense of value of teachers, this paper probes into the concrete implementation path of information technology promoting ideological and political education in classroom, on-line and off-line.

1 Logical basis of the integration of information technology and teaching in ideological and political education

The logical framework of high integration of ideological and political course teaching and information technology emphasizes the research and development and sharing of digital high-quality educational resources with the network as the carrier, focusing on the exploration of the teaching mode in the information age, to realize the functions of teaching, learning and communication in and out of class, and to explore the rational allocation of teachers resources and the direction of Teachers' Professional Development. The main contents of its logical framework include the construction of a teaching system based on problem logic, the research and development of high-quality teaching resources for ideological and political courses, and the construction of a data integration center system supported by data mining technology, construct the interactive teaching mode of teachers and students under the support of network platform, and create the teaching team of the ideological and political course integration of full-time and part-time based on the connecting mode.

Modern information technology in promoting the emergence of new educational models, creating a new learning environment, promoting the reform and innovation of teaching methods, etc. , it provides theoretical guidance and practical basis for the integration of ideological and political education and information technology. The development of the open education movement and the teaching mode of the media-oriented university have brought a strong driving force to the high integration of information technology and ideological and political education. The World Open Education Resources movement has accelerated the reform of the education system and the innovation of its teaching and learning methods. Large-scale online courses not only provide free learning resources, but also realize the concept of full participation in the teaching process, the core of which is to emphasize "learning", that is, student learning. The transformation from the traditional teacher-dominated classroom to the student-centered classroom reflects the role-changing relationship between teachers and students in the teaching structure and process.

The survival characteristics and dissemination mode of modern information technology coincide with the psychological development and growth characteristics of contemporary college students, which provides a new space and thinking for the high integration of ideological and political education teaching and information technology. As college students grow up under the condition of contemporary information, the survival characteristics and environment of modern information technology, which are both virtual and realistic, coincide with the cognitive psychological development characteristics of contemporary college students, this provides a new development space for ideological and political education. This space meets the needs of contemporary college students to express their emotions, and provides a new development space for the high integration of information technology and ideological and political education.

2 The dilemma of ideological and political education with the integration of information technology and teaching

The trend of deep integration of it and ideological and political theory teaching is irresistible. The ideological and political theory teaching should actively adapt to the development of it. However, from the current point of view, the teaching of information technology and ideological and political theory courses has not really achieved deep integration, due to the lack of research on the rules and laws of the deep integration of information technology and ideological and political theory courses, etc. , lack of a dynamic perspective on the ideological and political theory teaching and information technology in-depth integration of the system and mechanism.

The importance of information technology to teaching is self-evident, but it is not advisable to pay too much attention to the diversification of information technology and the high-end of technical equipment. The extensive use of information technology does not mean the realization of teaching objectives, nor does it mean the inevitable improvement of teaching quality, it is important to find information technology and technical equipment suitable for your school and students.

Some teachers, when making multimedia courseware, pay attention to the collection of materials, the beautification of pictures and the increase of forms. They lack a careful classification and induction of students' points of concern, interest and knowledge to be mastered, make it difficult for students to grasp the theme, theme and focus of the course. Some teachers have made new designs in their teaching courseware. Words, pictures, videos, animations, sounds and so on are dizzying, even dazzling. Students' attention to study is easily diverted, not only difficult to concentrate on learning knowledge, but also the lack of emotional communication with teachers, communication opportunities and time. In some ideological and political courses, the teaching process is relatively compact, and there is no time and opportunity for teachers and students to communicate at all. The advantages that face-to-face teaching should have are ignored. It is of course a great advantage for face-to-face teaching to control the classroom, but we can't ignore the emotional connection between teachers and students.

Although there are differences in methods and time and so on, teaching inspection is done in almost every school. In some schools, the inspection of teaching can not be described as careless. The syllabus, the examination syllabus, the teaching plan, the student roster, the courseware, the preparation of textbooks, the preparation of materials for classes, and so on, have all been checked in place, there is no shortage of on-site lectures, of course, but the problem is that these materials are still some distance from the real organization of the classroom. Teachers have different organizational ability, and the classroom effect is very different.

The high integration of information technology and ideological and political education teaching is not the application of technology in general sense, but the application of information technology in ideological and political education teaching, promote the reform of teaching idea, teaching mode and learning mode, realize the integration and mutual promotion of information technology and ideological and political education. Ideological and political course teaching and information technology are highly integrated, not only refers to the possession of modern information technology and the use of some information technology means in the process of education and teaching, but also lies in: On the one hand, it is necessary to integrate information technology into the whole process of ideological and political education and teaching, change the traditional education and teaching mode, and give birth to new teaching methods and modes, on the other hand, if the ideological and political course is to practice the new educational and teaching ideas and models, it must be supported by the corresponding information technology, at the same time, it also provides a new idea and direction for the development of information technology. This is the ideological and political education teaching and information technology high integration of the basic requirements.

3 The construction of ideological and political education teaching mode based on the integration of information technology and teaching

The basic idea of high integration of ideological and political education and information technology is to adopt the application-driven strategy. Only by focusing on the application of information technology in the teaching process can we gradually realize two basic changes in the teaching mode of ideological and political education Second, the ability-based teaching mode to the values of training-based teaching mode change. Only in this way can we really discover the deep fusion point between ideological and political education and information technology, it embodies the strong leading and supporting function of high integration in promoting the reform and development of ideological and political course teaching and improving the quality and level of ideological and political course teaching. High integration includes three aspects: first, the application of modern information technology in ideological and political education teaching process, that is, in the main battlefield of teaching and learning, especially in the classroom teaching process. This application must focus on solving the problems in the reform and development of ideological and political education, providing high-quality learning experience for college students, and solving the problems of students' ideological understanding. Second, the modern information technology in the classroom teaching process in-depth universal application. Starting from the goal requirements of ideological and political education and the needs of college students, the majority of teachers and students want to use, love to use, used to use, it has become a normal application of the regular teaching and learning activities that teachers and students widely participate in, and has become a habit and normal state of teaching and learning. Third, modern information technology in the ideological and political education teaching reform and development process constantly explore new applications, try new applications, to achieve new applications. It is foreseeable that the continuous innovation of modern information technology will provide new sources and impetus for new applications, and the needs and development of ideological and political education will also put forward new requirements for the development of modern information technology. It is necessary to promote the construction by application, to reform and upgrade the existing teaching methods, and to focus on solving the major theoretical and practical problems in the reform and development of ideological and political courses.

At present, the information technology in the field of vision that ideological and political education and information technology are highly integrated mainly includes two aspects: tools, that is, various available technologies (or information media) , such as computer and its technology, multimedia and its technology, network communication and its technology, database and its technology, etc. . Method is an intelligent form of technology, including information processing and information dissemination technology, that is, the collection of information, processing, storage, dissemination and application of technical methods. Computer Technology is the core of modern information technology, multimedia technology and network communication technology are the focus of current information technology. High integration refers to the integration of information technology with multimedia technology and network technology into the teaching objectives, teaching methods, teaching preparation, teaching process and teaching feedback of ideological and political theory courses, creating a new type of teaching environment, making the classroom more vivid and flexible, teaching more targeted, learning methods more convenient, learning better results, so as to give full play to the enthusiasm of teachers and students' initiative in learning, to realize the successful transformation from teaching material system to teaching system and knowledge system to belief system.

The deep integration of ideological and political education teaching and information technology needs ideological and political teachers to strengthen the theoretical study of information technology, deeply understand the connotation and significance of "deep integration", and integrate information technology into the teaching objective of ideological and political education, run through the ideological and political teaching process, thus creating a new classroom teaching environment, stimulate students' willingness to learn independently, play the enthusiasm of teachers teaching. Information Technology, with computer, multimedia technology and network communication technology as its core, is an important prerequisite for carrying out high-quality and high-efficiency teaching. Teachers must constantly update their knowledge structure to meet the needs of the times, in order to make ideological and political education teaching more vivid, flexible and effective.

In addition to the teaching software commonly used by teachers, the ideological and political education teaching needs to develop more professional and more targeted software. First, professional teaching requires professional software. Secondly, there are ideological and political courses at the stage of university education, involving a large number of course subjects and teachers, the software development, the application will be more extensive. The development of special teaching software should follow the following principles: firstly, the knowledge of ideological and political course must be embodied, and the main contents that should be grasped in all ideological and political course should be embodied in the software, or it should be convenient to input and search, this is the most basic requirement. Second, it must be artistic. Politics classes are full of the arts of study, work and life. If the relevant software can not reflect this, then such software has no depth. Finally, it must be ideological. The ideological and political course is full of philosophy and historical experiences and lessons. It needs teachers and students to constantly think and look for them, must reflect the ideological, which should be ideological and political course software and other software differences between the watershed.

Strengthen the actual effect of ideological and political education by means of information technology. Network ideological and political work, as an extension of traditional ideological and political work, has changed the one-to-one, face-to-face communication and communication model of ideological and political education in the past, to quickly understand the ideological situation and practical needs of college students, so as to facilitate teachers to accurately grasp the ideological pulse of college students and guide the unhealthy tendencies and extreme views in a timely manner, improving the actual effect of ideological and political education in an all-round way. In the work of ideological and political work on the Internet, we will use information technology to create a platform for the study and evaluation of the wisdom of public opinion on the Internet, submit regular special reports on public opinion to the school, give timely feedback on the ideological trends of teachers and students in the school, lead the trend of public opinion, and consolidate the ideological foundation, take the high ground. Through the network public opinion information staff to do a good job of information release and guidance, to play the role of the uplink and uplink, enhance the effectiveness of ideological and political education.

4 Conclusion

The deep integration of ideological and political education and information technology needs the process of "practice-feedback-re-practice". The fusion product of information technology and ideological and political education, such as "Moke", is a brand-new teaching mode, which has been introduced into China for only a few years, it is inevitable that some new problems will appear in the process of practice. At present, there are not many universities offering "MOOC" independently in China. There are many problems in ideological and political teaching under the MOOC mode. We should follow up the problems arising in the process of integration in time, gET regular feedback from students and teachers on current issues and complement each other with a combination of information technology and traditional classroom instruction.

References

1. Li Liang. The idea and concept of the high integration of ideological and political education and information technology [J]. China Higher Education, 2017, (24): 20-22.
2. Fang Yanrui. Exploration on the Integration of Information Technology and Ideological and Political Course Teaching in Colleges and Universities [J]. Information Recording Materials, 2017, (11): 127-128.
3. Tan Yuhan; Zhang Huaimin. Exploration on Teaching Affinity of Ideological and Political Theory Courses in the Modern Information Technology Environment [J].

School Party Building and Ideological Education, 2017, (7): 50-51+56.
4. Zhao Qingsi. The alienation and transcendence of the deep integration of modern information technology and ideological and political courses in colleges and universities [J]. Academic Forum, 2018, (5): 162-167.
5. Mao Rui. Feasibility and Countermeasures of the Integration of Modern Information Technology and Ideological and Political Course Teaching in Colleges and Universities [J]. Western Quality Education, 2018, (21): 107-108.

Research on Financial Market Risks Based on VaR Model

Rongming Wang

Oxbridge College, Kunming University of Science and Technology, Kunming, Yunnan, 650106, China

Abstract. With the continuous development of the Internet financial industry, its impact on people's daily life is growing. Under the influence of the information technology and the financial innovation, the volatility of the financial market has been significantly enhanced, and the financial risk has become increasingly serious. The COVID-19 in 2020 still has a profound impact on the global financial market. The reasonable measurement of the financial risks is the most important step in the financial risk management. Based on the sample analysis of the Standard & Poor's 500 Index from January 3, 2013 to June 30, 2020, this paper measures the risks by analyzing and calculating the value at risk (VaR) of the sample data by different methods. This paper mainly establishes the ARMA model, the GARCH model based on bayese statistics, and the POT model based on the extreme value theory, and compares and analyzes the VaR values obtained by the three models. Finally, the effectiveness of VaR is tested by VaR backtracking test, and the advantages and disadvantages of the methods are compared and evaluated.

Keywords: VaR, ARMA model, GARCH model, extreme value theory, POT model, R, openbugs.

1 Introduction

Since the 21st century, human beings have entered the era of the Internet economy, and governments around the world have slowed down the regulation of the financial industry, and the degree of the financial liberalization has been strengthened. Under the influence of the information technology and the financial innovation, the volatility of the financial market has been significantly enhanced, and the stability of the financial system has declined. As a result, the financial risks of the financial institutions, industrial and commercial enterprises, residents and even the country have become increasingly serious. Financial risks not only affect the normal operation of the financial institutions and industrial and commercial enterprises, but also pose a serious threat to the financial and economic stability of the country and even the world. The frequent occurrence of the financial crisis has brought a series of serious consequences.

Since 2020, the COVID-19 has swept the world, and profound changes have taken place in the world economic situation, and the fluctuation of the global financial market has intensified. Facing the impact of the epidemic, major economies launched a new round of the monetary easing policies. At present, the interest rates in the United States, the United

Kingdom, Canada and Australia are close to zero, and the total economic volume of the regions with negative interest rate policies such as the euro zone and Japan has exceeded 20% of the global GDP. The extremely loose monetary environment has changed investors' expectations and intensified the financial market volatility. Since 2020, the US dollar index has fluctuated significantly, and commodity prices have been adjusted in depth. The global debt scale has exceeded $260 trillion, a record high, and the stock markets of major economies have fluctuated at a high level, which seriously deviates from the weak performance of the real economy. In view of the wide and far-reaching influence of the financial risks, the financial circles, business circles, government authorities and international financial organizations all over the world pay close attention to and attach great importance to it. Countries and major financial institutions have also issued new policies, which put forward strict requirements for the risk management of financial institutions.

The movement of the financial market has produced a large number of data. How to get a quantitative index from this large amount of data to reasonably measure the market risk and measure the financial risk is an important part of the financial risk management. Underestimating risks can cause huge problems (such as bankruptcy) for banks and other parts of the financial markets. On the other hand, overestimation of risks may make investors too sensitive to risks and miss the investment opportunities, thus reducing the return. Therefore, it is very important to estimate the risks correctly. Through the continuous research of many scholars, there are many methods to measure the financial risks. Among them, the Value at Risk (VaR) method is very popular because of its intuitive, concise and scientific nature. It can integrate the market risks of different trading and business departments into a numerical value, so it is easy to understand and manage. VaR method can not only clearly describe the size of the market risks, but also has the support of rigorous probability and statistics theory. At present, it has become the most extensive and important method in the modern risk quantitative management. VaR is an analytical measure that has several advantages. It is easy to estimate and easy to understand for non economics professionals. The precise horizontal two-week to 10-day VaR estimates of 99% quantiles are accepted by the Basel Committee for banks and the internal clearing house of regulatory banks (Basel Committee on Banking Supervision, 1995). This stems from concerns that financial institutions may not be able to liquidate their two-week liquidity crisis.

1.1 The definition of VaR

Jorion (1996) gave the most authoritative definition of VaR: "VaR is to measure the possible or potential loss of the financial assets or portfolio under the future asset price fluctuations. Under the normal market conditions and the given confidence level, the worst expected loss value of a financial asset or portfolio of assets in a future holding period."
The mathematical expression of VaR is as follows:

$$P(W_{t+H} - W_t < -VaR_{W,t}(H,\alpha)) = \alpha \qquad (1)$$

In it, W_t is the value of the asset at time t, and $VaR_{W,t}(H,\alpha)$ is the VaR of the holding period of H of the financial asset at the confidence level $1-\alpha$ (or significant level α), which indicates that in a future holding period H, the probability of the financial asset loss exceeding $VaR_{W,t}(H,\alpha)$ is α.

VaR: We can also use the yield of the financial assets to calculate VaR:

$$P(r_{t+H} < -VaR_{W,t}(H,\alpha)) = \alpha \qquad (2)$$

In it, $r_{r+H} = \ln(W_{t+H}) - \ln(W_t)$, and r_{r+H} is the continuous compound rate of return. Furthermore, if the yield distribution of financial assets is regarded as a random variable, then VaR is a specified quantile of the probability distribution of the return. Assuming that the known r_{r+H} probability distribution density function is $f(x)$, then the VaR of the financial asset yield at the confidence level $1-\alpha$ is the lower α quantile of the distribution $f(x)$. That is:

$$\int_{-\infty}^{-VaR_{W,t}(H,\alpha)} f(x)dx = \alpha \tag{3}$$

As a measure of risks, VaR's calculation depends on the time interval t, the confidence level $1-\alpha$ and the density function $f(x)$ of the expected return rate.

1.2 VaR research methods

Before calculating VaR, we must have a certain understanding of the financial market data, and on this basis, we can correctly model and measure VaR. Since the researches of Mandelbrot (1963) and Fama (1965), it has been found that the financial market data has the following characteristics:

(1) The distribution of the financial yield is high peak and thick tail. Compared with the normal distribution, it has thicker tail and higher peak value.

(2) The yield is a typical negative skew.

(3) The square of the return rate has the typical autocorrelation. That is, the volatility of market factors tends to agglomerate.

All the traditional VaR models are based on these empirical laws.

Theoretically, VaR research methods are divided into three categories [defined by McNeil and Frey 92000, P. 272)]:

Parametric type: modeling, research and calculation based on the sample data, such as RiskMetrics and GARCH models.

Semi parametric type: it is mainly used for modeling, research and calculation of the sample tail data, such as the extreme value theory and CAViaR.

Nonparametric: that is, through the simulation research and calculation, such as the historical simulation and Monte Carlo simulation.

1.3 Features of this paper

This paper selects the daily closing price of the Standard & Poor's 500 Index from January 3, 2013 to June 30, 2020 as the sample for the statistical modeling and analysis. The data volume is relatively large, which can avoid the impact of the insufficient sample size on the estimation of the model. Secondly, this paper selects VaR as the index to measure the financial risks, which has its corresponding theoretical basis as the support. At the same time, the definition of the VaR index is more popular and can be accepted by most people. In addition, this paper is not limited to one method, but involves two kinds of methods, which cover a wide range. For the estimation of the parametric model, this paper not only uses the most common maximum likelihood estimation method, but also uses the bayese statistical method to estimate the parameters of the model by making full use of the information of the data itself. Firstly, the ARMA (1, 1) model is established to find the function model which can fully reflect the rate of return. Then, based on the ARMA (1,1) model, considering the conditional autoregression of data, the GARCH (1,1) model is established. For the semi parametric model, the pot model in the extreme value theory is selected for modeling, and the sample data is analyzed by analyzing the tail data exceeding the threshold value. Finally, the VaR values calculated by the three models are compared and analyzed, and reasonable explanations are given and the results are further analyzed and discussed. And the calculated value of VaR is tested retrospectively to verify its accuracy.

2 Data preprocessing

Since China's stock market has only 20 years of history, the regularity of the market data is not easy to find. Therefore, this paper selects 1825 data of the daily closing price of the Standard & Poor's 500 Index from January 3, 2013 to June 30, 2020 as samples for analysis.

In 1957, the Standard & Poor's 500 Index, which is a large-scale free trading index, was issued in 1957. The stocks included in the S & P 500 Index are shares of the large public companies traded on the New York Stock Exchange and Nasdaq, the largest stock exchange in the United States. The index focuses mainly on companies in the United States, but there are also some top companies or multinational companies in other countries.

The S & P 500 Index is the most widely referenced stock index after the Dow Jones Industrial Average (DJIA). It is considered to be the leader of the US economy and is included in the leading economic indicators index. Many mutual funds, exchange traded funds and other funds, such as pension funds, aim to follow the trajectory of the S & P 500 Index. Hundreds of billions of dollars have been invested in this way.

McGraw Hill called the index the best of many indicators owned and maintained by S & P. The Standard & Poor's 500 Index is not only an index, but also the common stock index of 500 companies included in the index. There are many codes for the S & P 500 Index. For example, "GSPC," "INX," and "$SPX". Stocks covered in the S & P 500 Index are also included in the S & P 1500 and S & P Global 1200 stock indexes.

First of all, we make an intuitive trend chart of the closing price of the stock.

From Figure 1, we can roughly see the daily closing price trend of the Standard & Poor's 500 Index from January 3, 2013 to June 30, 2020. From the chart, we can find that at a certain stage, the closing price has a rapid decline, which happens to be in 2020. In fact, during this period, the whole world is in the financial crisis, and the daily closing price of stocks can clearly show this change.

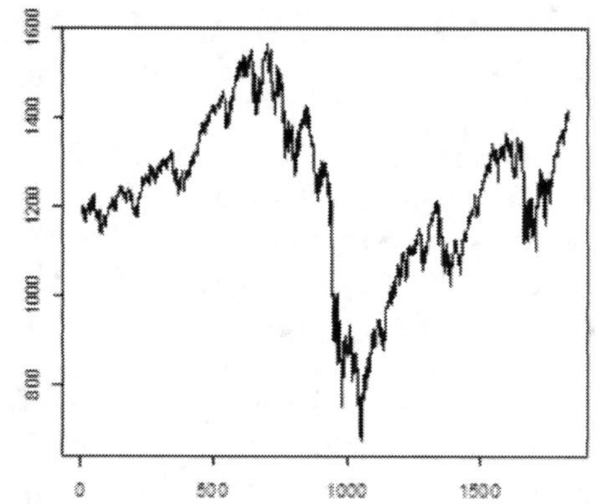

Fig. 1. Line chart of closing prices on sample day.

It can be seen from Figure 1 that the sample data is not stable, and what this paper studies is the return rate of the stock. Therefore, it is necessary to convert the sample data into a stationary time series. The yield at the time of return is defined as R_t,

$R_t = \ln(P_t / P_{t-1})$, where P_t is the closing price of the time t.

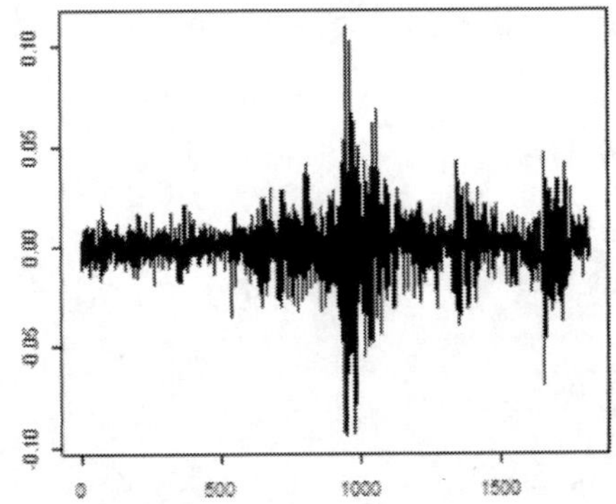

Fig. 2. Line chart of the yield.

It can be seen from Figure 2 that the transformed yield data conforms to the stationary stochastic process. After a preliminary descriptive statistical analysis of the data, Table 1 can be obtained.

Table 1. Descriptive statistical results of yield.

	N	Minimum value	Maximum value	Mean value	Standard deviation	Skewness	Kurtosis
R	1824	-0.0947	0.1096	0.0001	0.0144	-0.2979	9.2677

From the skewness value in Table 1, it can be seen that the yield data has obvious left skewness, and the kurtosis value is obviously greater than 3. It can be concluded that the data has obvious peak characteristics, which conforms to the high peak and thick tail characteristics of the financial data.

3 Model one: ARMA model

3.1 Model concept and principle

The financial time series data generally have autocorrelation effect and lag effect, so the ARMA model is often used for modeling and analysis. The ARMA model is sometimes called the Box-Jenkins model. Given a set of time series data X_t, ARMA model is used to understand or predict the future value of this group of time series. ARMA model consists of two parts: autoregression (AR) and moving average (MA). The model is usually called ARMA (p, q) model, where p is the lag order of the autoregressive part and q is the lag order of the moving average part.

AR (p) stands for the autoregressive model with p-order lag. AR (p) model was written as follows:

$$X_t = \alpha_0 + \alpha_1 X_{t-1} + \alpha_2 X_{t-2} + ... + \alpha_p X_{t-p} + \varepsilon_t \quad (4)$$

$\alpha_1, ..., \alpha_p$ is the parameter of the model, and α_0 is the constant, while ε_t is the white noise. For simplicity, many people ignore the constant term.

In order to keep the model stable, some restrictions should be made on the parameter values.

MA (q) represents the autoregressive model with q-order lag. MA (q) model was written as follows:

$$X_t = \mu + \beta_1 \varepsilon_{t-1} + \beta_2 \varepsilon_{t-2} + \ldots + \beta_q \varepsilon_{t-q} + \varepsilon_t \quad (5)$$

In it, $\beta_1, \ldots, \beta_q$ is the parameter of the model, and μ is the expected value of X_t (often assumed to be 0), and $\varepsilon_t, \varepsilon_{t-1}, \ldots$ is the white noise error.

Therefore, the formula of ARMA (p, q) model is as follows:

$$X_t = c + \varepsilon_t + \sum_{i=1}^{p} \alpha_i X_{t-i} + \sum_{i=1}^{q} \beta_i u_{t-i} \quad (6)$$

3.2 Establishment of the model

The most important step in the establishment of ARMA (p, q) model is to select the sum of lag orders p and q, and first draw the autocorrelation graph (ACF) and the partial autocorrelation graph (PACF).

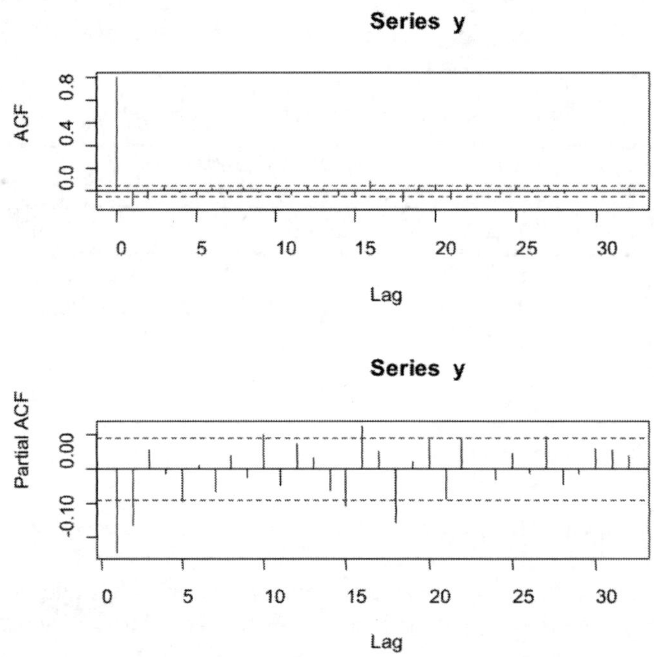

Fig. 3. Data autocorrelation and partial autocorrelation graph.

It can be seen from Figure 3 that the second-order autocorrelation and the partial autocorrelation of the data are significant. Therefore, it can be preliminarily judged that $p \leq 2, q \leq 2$. The final model is ARMA (1,1) model by comparing the AIC of the model and the significance of the model variables when taking different values. The results of parameter estimation are as follows:

arima(x = y, order = c(1, 0, 1))

Coefficients:

	ar1	ma1	intercept
	0.2720	-0.4075	1e-04
s.e.	0.1437	0.1364	3e-04

sigma^2 estimated as 0.0002036: log likelihood = 5163.48, aic = -10318.96

It can be concluded that the final model is as follows:

$$y_t = 0.2720 * y_{t-1} - 0.4075 e_{t-1} + \mu_t \quad (7)$$

The residual error of the model is tested, and the result is

Box-Ljung test

data: model$residuals

X-squared = 12.1953, df = 8, p-value = 0.1427

The original hypothesis can not be rejected, which means that the residual conforms to the white noise.

4 Model 2: GARCH (1, 1) model based on the Bayesian estimation

4.1 Model concept and principle

The traditional ARMA (p, q) model assumes that the fluctuation amplitude (variance) of time series variables is fixed, but in fact, this is not in line with the reality, which makes the traditional time series analysis ineffective for practical problems. Robert Engle (1982) proposed the arch model, which successfully solved the volatility problem of time series.
In the field of economics, the Auto Regressive Conditional Heteroscedasticity (ARCH) model is used to analyze and simulate a group of time series. This model is used whenever there is reason to believe that there is a specific size or variance at any point in the sequence. The ARCH model assumes that the variance of the current or improved error term is a function of the actual value of the error term in a previous period, and the variance is usually related to the square of the previous improved error term.

Let's simulate a time series using the ARCH process. e_t is the error term (the residual return value of the mean equation). It can be divided into a random factor ε_t and a time-based standard deviation $\sqrt{h_t}$:

$$e_t = \varepsilon_t \sqrt{h_t} \quad (8)$$

The random variable z_t is a strong white noise process, and the model of h_t is:

$$h_t = \alpha_0 + \alpha_1 e_{t-1}^2 + \ldots + \alpha_q e_{t-q}^2 = \alpha_0 + \sum_{i=1}^{q} \alpha_i e_{t-i}^2 \quad (9)$$

In it, $\alpha_0 > 0, \alpha_i > 0, i > 0$.

ARCH (q) model can be estimated by the least square method.

If the variance is represented by the autoregressive moving average model (ARMA model), then the model is a generalized autoregressive conditional heteroscedasticity model, namely the GARCH model [Bollerslev (1986)].

The model can be expressed as follows:

$$\begin{cases} e_t = \varepsilon_t \sqrt{h_t} \\ \varepsilon_t \mid F_{t-1} \sim iid(0,1) \\ h_t = \beta_0 + \beta_1 e_{t-1}^2 + \ldots + \beta_p e_{t-p}^2 + \gamma_t h_{t-1} + \ldots + \gamma_q h_{t-q} \end{cases} \quad (10)$$

In fact, because:

$$e_t = y_t - \alpha_0 - \alpha_1 y_{t-1} - \alpha_2 y_{t-2} - \ldots - \alpha_p y_{t-p} \quad (11)$$

The variance equation is related to the residual term of lag, so the parameter estimation of the variance equation should be combined with the parameter of the mean equation.

For the determination of the GARCH model parameters, this paper chooses the Gibbs sampling based on the bayese statistics.

Bayesian statistics is a systematic statistical inference method named after Thomas Bayese (1763). Later generations extended and developed the bayese statistics on his basis, and eventually formed an influential school, namely the Bayesian School. Different from the classical statistics, the bayese statistics has its own systematic theoretical framework. The basic view of the bayese school is that in any statistical inference problem θ, and in addition to using the information provided by sample $X=\{x_1,...,x_n\}$, a prior distribution must be determined in advance as an indispensable element θ in the statistical inference. The prior distribution can be arbitrarily determined subjectively. According to the distribution $p(x_1,...,x_n\mid\theta)$ of sample X and the prior distribution $\pi(\theta)$ of θ, under known condition $X=x$, the conditional distribution $\pi(\theta\mid x_1,...,x_n)=\dfrac{p(x_1,...,x_n\mid\theta)*\pi(\theta)}{p(x_1,...,x_n)}$ obtained of θ is called posterior distribution. According to the posterior distribution, the statistical inference analysis was carried out.

Gibbs sampling method is a Markov chain Monte Carlo simulation (MCMC) algorithm, which is used to obtain a series of the random samples with multiple distributions. Since the direct sampling is difficult, this process can roughly estimate the joint distribution. The basic principle of the Gibbs sampling is: suppose we want to get k groups of samples $X=\{x_1,...,x_n\}$ from a joint distribution $p(x_1,...,x_n)$, and assume that the i sample is $X^{(i)}=\{x_1^{(i)},...,x_n^{(i)}\}$. The sampling process was as follows:

(1) We assume an initial value for all variables $X^{(0)}$

(2) The sampling of each variable $x_j^{(i)}$ in the $i=\{1...k\}$ sample is based on conditional distribution $p(x_j^{(i)}\mid x_1^{(i)},...,x_{j-1}^{(i)},x_{j+1}^{(i-1)},...,x_n^{(i-1)})$. In other words, the sampling of each variable is based on the conditional distribution of other samples. After sampling, it is updated to the latest sampling value.

4.2 Establishment of the model

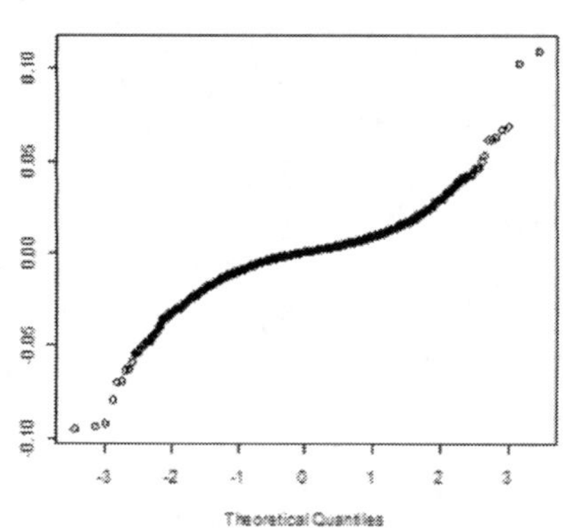

Fig. 4. Normal distribution QQ chart.

From the QQ chart of the data, we can see that the data does not conform to the normal distribution, and has obvious characteristics of sharp front and thick tail, so it is not suitable to use the GARCH model under the normal distribution to fit the data. Considering that t distribution is a common thick tailed distribution, the GARCH model of t distribution is used for fitting. The specific model is as follows:

$$\begin{cases} e_t = \varepsilon_t \sqrt{h_t} \\ \varepsilon_t \mid F_{t-1} \sim t(0,1,v) \\ h_t = \beta_0 + \beta_1 e_{t-1}^2 + \gamma_1 h_{t-1} \end{cases} \tag{12}$$

As for the prior distribution selection of parameters, the prior distribution of v degree of freedom is more suitable for $1/(1+v)^2$, and the prior distribution of other parameters is the uniform distribution of upper $[0,1]$ without information.

Let the observed value be $Y = (y_1, y_2, \ldots, y_n)$, and then the likelihood function of the parameter is:

$$I(\theta \mid y) \propto \prod_{t=1}^{T} \frac{\Gamma(\frac{v+1}{2})}{\Gamma(\frac{v}{2})} (vh_t)^{-\frac{1}{2}} [1 + \frac{y_t^2}{vh_t}]^{-\frac{v+1}{2}} \tag{13}$$

The full distribution conditions are as follows:

$$v \sim \prod_{t=1}^{T} \frac{\Gamma(\frac{v+1}{2})}{\Gamma(\frac{v}{2})} (vh_t)^{-\frac{1}{2}} [1 + \frac{y_t^2}{vh_t}]^{-\frac{v+1}{2}} (1+v^2)^{-1} \tag{14}$$

$$\alpha_0 \sim \prod_{t=1}^{T} \frac{\Gamma(\frac{v+1}{2})}{\Gamma(\frac{v}{2})} (vh_t)^{-\frac{1}{2}} [1 + \frac{y_t^2}{vh_t}]^{-\frac{v+1}{2}} \tag{15}$$

The full conditional distribution of $\alpha_1, \ldots, \alpha_p, \beta_1, \ldots, \beta_q$ is the same.

For the data in this paper, the simplest GARCH (1, 1) model is selected for modeling.

For the parameter estimation of GARCH (1, 1) model, this paper adopts the Gibbs sampling method based on the bayese statistics.

In the simulation process, 1000 Gibbs pre iterations are performed to ensure the convergence of the parameters. Then the original pre iteration is discarded and 10000 iterations are performed.

This paper implements the Gibbs sampling process through the Openbugs program.

The sampling results are shown in Table 2:

Table 2. Openbugs results (see Appendix for other simulation results).

	mean	sd	MC_error	val2.5pc	median	val97.5pc
alpha0	-8.99E-05	0.004326	5.87E-05	-0.008634	-6.07E-05	0.008647
alpha1	-0.01418	0.577	0.04689	-1.12	0.008043	1.032
beta0	39.69	4.898	0.04352	32.61	37.88	48.3
beta1	0.5004	0.2898	0.00366	0.02276	0.4972	0.9804

gama	0.9963	0.003754	8.66E-05	0.9863	0.9974	0.9999

According to the results in Table 2, the model is as follows:

$$\begin{aligned} e_t &= y_t + 8{,}99E-5+0.014y_{t-1} \\ h_t &= 39.69+0.5004e_{t-1}^2+0.9963h_{t-1} \end{aligned} \quad (16)$$

From the results of Table 2, it can be found that the parameters of each variable of the model are significant at 95% confidence level.

5 Model 3: POT model of the extreme value theory

5.1 Model concept and principle

Extreme value theory or extreme value analysis (EVA) is a branch of statistics that deals with the extreme deviation of the median of probability distribution. It aims to assess the probability of an ordered sample of a given random variable to occur more extreme than any previously observed event. Extreme value theory is widely used in many fields, including structural engineering, finance, geography, traffic prediction and geological engineering.

In this paper, we mainly use the Peaks over Threshold (POT) method to analyze the yield.

Let $X_1, X_2, ..., X_n$ is an independent and identically distributed sample from the asset income population X, and the distribution function is $F(x)$, and select a threshold value u. If $X > u$, it is called over threshold, and $Y = X - u$ is also called the over-threshold value. According to the conditional probability formula, the distribution function of Y is as follows:

$$F_u(y) = p\{X-u \le y \mid X > u\} = \frac{P(y+u)-F(u)}{1-F(u)} \quad (17)$$

Thus, we can get the following results:

$$F(x) = F(y+u) = [1-F(u)]F_u(y)+F(u) \quad (18)$$

The key to the solution to $F(x)$ is the expression of $F_u(y)$ to be obtained. According to Pickandes Balkemade Haan theorem, for a sufficiently large threshold u, the over threshold distribution $F_u(y)$ can be approximated by the generalized Pareto distribution, whose expression is as follows:

$$G_{\xi,\beta}(y) = \begin{cases} 1-(1+\xi * \frac{y}{\beta})^{-1/\xi}, \xi \ne 0 \\ 1-\exp(-y/\beta), \xi = 0 \end{cases} \quad (19)$$

In it, $y = x - u$.

If $\xi \ge 0$, $y \ge 0$, but if $\xi < 0$, $0 \le y \le -\beta/\xi$, where ξ and $\beta(\beta > 0)$ are the shape parameters and scale parameters of the generalized Pareto distribution. If $\xi > 0$, the distribution function was ordinary Pareto distribution, which was a kind of thick tailed distribution, and the financial event sequence generally belonged to this type. If $\xi = 0$, the distribution function was the exponential distribution. If $\xi < 0$, it was Pareto distribution, which was a thin tail distribution.

5.2 Establishment of the model

In order to establish the POT model, we need to determine the threshold value first. In order to make the determination more scientific, we use the combination of the mean residual life diagram and the Hill chart to observe.

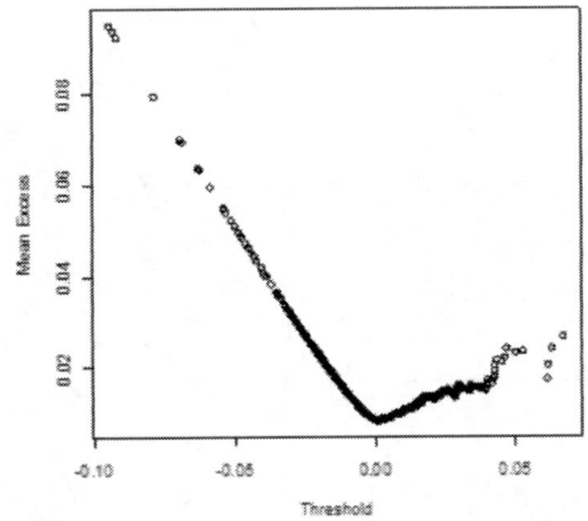

Fig. 5. Average remaining life chart.

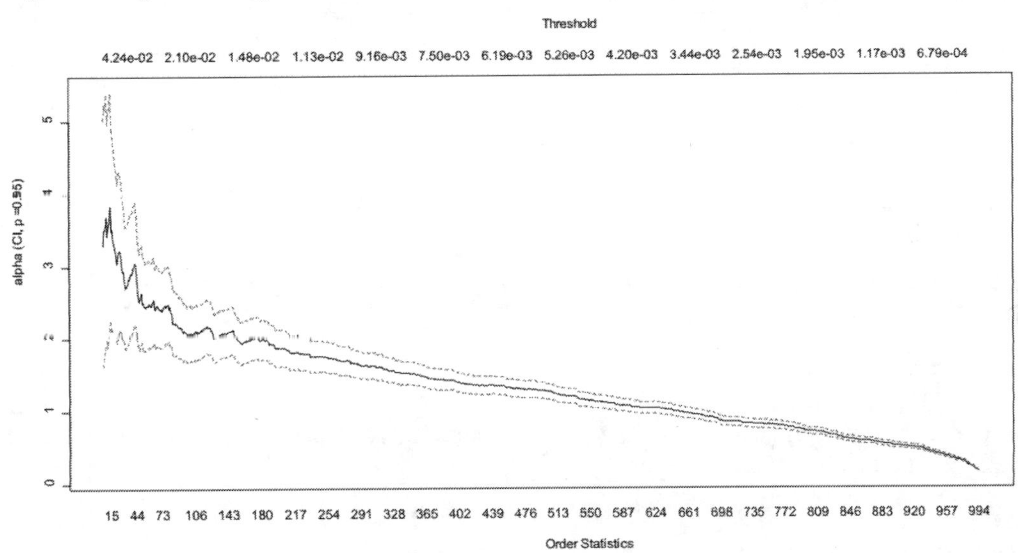

Fig. 6. Hill chart.

It can be seen from Figures 5 and 6 that the selected threshold value is 0.0113 and the number of overruns is 230. The maximum likelihood method is used to estimate the data. Get the final result $\xi = 0.2119, \beta = 0.0085$.

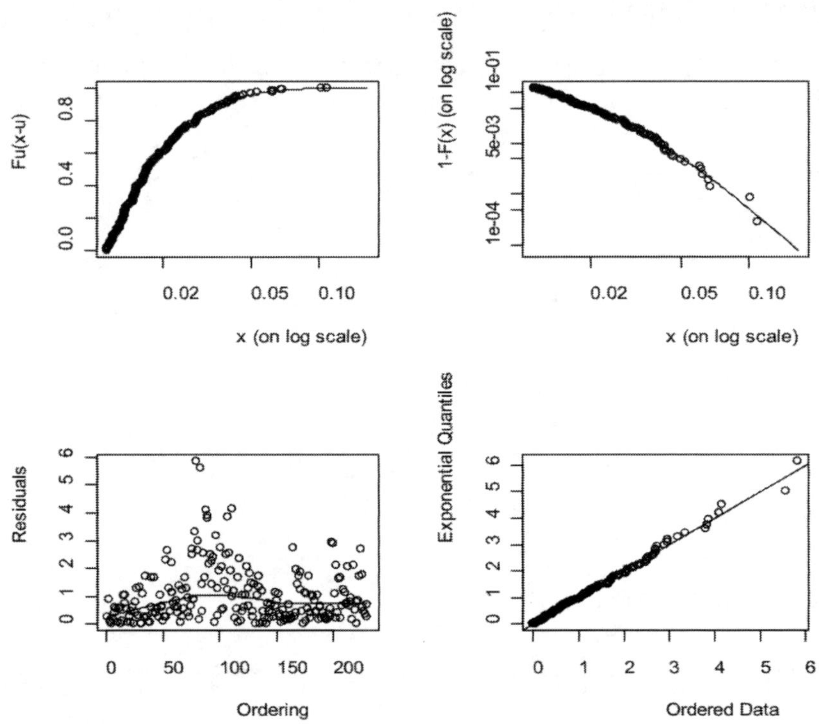

Fig. 7. GPD estimation results.

It can be seen from Figure 7 that the fitting result of the model is very good.

6. Calculation of Value at Risk (VaR) under each model

6.1 Calculation of VaR

In retrospect, the definition of VaR F is the cumulative distribution function of future earnings, and then:

$$VaR_p = -F^{-1}(1-p) \tag{20}$$

Therefore, VaR is essentially a quantile of future earnings.

6.1.1 Parameter method

The parameter method is to estimate the parameters of the distribution on the assumption that the future income satisfies a certain distribution (usually normal distribution), so as to determine the whole distribution and then calculate the quantile of the distribution.

It is reasonable to assume that the future returns satisfy the normal distribution:

(1) The short-term performance of stock returns can be approximated by the joint normal distribution;

(2) Most assets can be expressed as a linear combination of risk factors;

(3) The linear combination of any number of the normal distributions is still the normal distribution, so the linear expected return distribution of a portfolio is still the normal distribution, and its distribution parameters are unique and fixed.

Taking Model 2 as an example, the estimated value of unknown parameters is obtained by calculating the likelihood function of the GARCH model, and then the variance between the actual value and the estimated value of the model is calculated, that is, the square of the stock's ex ante volatility. According to the obtained volatility, the corresponding VaR is calculated. For example, under the 95% confidence level, the normal quantile can be multiplied by the estimated volatility.

6.1.2 Semiparametric method

Under the POT model of the extreme value theory, according to Formula 18, when the threshold value u is large enough, $F(u)$ can be estimated approximately by $\frac{n-n_u}{n}$, and n_u is the number of the over threshold samples, and n is the sample size. Most of the actual financial data show the characteristics of thick tail, that is, $\xi > 0$. It can be concluded that:

$$F(x)=1-\frac{n_u}{n}\left(1+\xi\frac{x-u}{\beta}\right)^{-1/\xi}, \xi>0 \quad (21)$$

Thus, the estimated value of VaR obtained by $F(VaR_p)=p$ under the fixed confidence level in extreme cases is given as follows:

$$VaR_p = u+\frac{\beta}{\xi}\left\{\left[\frac{n_u}{n(1-p)}\right]^{\xi}-1\right\}, \beta>0, \xi>0 \quad (22)$$

Therefore, according to the above three models, we can get three VaR estimates respectively.

Table 3. VaR values estimated by each model.

	VaR (95%) %	VaR (99%) %
ARMA	2.3472	3.3246
GARCH	2.3545	3.3349
POT	1.2127	1.3724

It can be seen from table 3 that the VaR values calculated by different model estimation methods are quite different. And according to the sample data, we can conclude that the VaR coverage of the results obtained by using the time series linear model is wider, and can reflect the risks of the samples better.

6.2 VaR backtracking test

If the estimated value of an asset does not fully take into account all the fluctuations affecting the valuation of the asset, the risk will be underestimated. In the day trading, the simulated profit and loss and VaR can pass the backtracking test, but the actual profit and loss and VaR can not pass the test. This is because the calculation of VaR is usually based on the daily closing price, and does not describe the volatility generated in the process of the day trading. In this case, in order to make the value of VaR reflect the risks of the asset profit and loss more accurately, a shorter time interval may be needed, or the VaR parameter may not be the most suitable risk description method, and other indicators should be used for the risk measurement.

The basic method of the backtracking test is to take VaR as the specific quantile of profit and loss, call the loss exceeding VaR as a "breakthrough" event, and test whether the "breakthrough" is an independent event with probability less than 5%.

In T samples, N is defined as the number of the VaR breakthrough events and N/T is the breakthrough rate. In the ideal case, N/T should tend to $p=0.05$ with the increase of the sample size. According to the requirements of the Basel framework (BIS), in order to make the detection more scientific, the sample data size should be at least one year.

Table 4. Basel framework requirements.

region	Number of breakthrough events n	Value added factor K
Green	0 —— 4	0
	5	0.4
	6	0.5
Yellow	7	0.65
	8	0.75
	9	0.85
Red	10+	1

There are clear requirements for the backtracking testing in the Basel framework. It limits only 99% of daily VaR and uses a one-year sample. The average number of breakthroughs in a year is about 2.5, and the Basel Committee can accept up to four breakthroughs. If there are five or more breakthroughs, the financial institutions will be in the yellow light or red light district, which requires higher capital of financial institutions, which is reflected in the increasing value-added factors.

Kupiec (1995) constructed a statistic to test the breakthrough rate of VaR:

$$LR_{uc} = -2\log((1-p)^{T-N}p^{N}) + 2\log((1-(N/T))^{T-N}(N/T)^{N}) \tag{23}$$

When T is large, the statistic accords with the χ^2 distribution of the degree of freedom of 1, so the corresponding hypothesis test can be carried out. If LR>3.84, the original hypothesis is rejected.

Table 5. Model back test, 95% confidence zone of non rejection test.

Probability level P	VaR confidence level	Non rejection area of failure number N		
		T=255 days	T=510 days	T=1000 days
0.01	99%	N<7	1<N<11	4<N<17
0.025	97.5%	2<N<12	6<N<21	15<N<36
0.05	95%	6<N<21	16<N<36	37<N<65
0.075	92.5%	11<N<28	27<N<51	59<N<92
0.1	90%	16<N<36	38<N<65	81<N<120

Data source: Kupiec (1995)

The S & P 500 Index from January 3, 2019 to December 30, 2019 is selected for calculation, and the results are as follows:

Table 6. VaR retrospective test table.

	VaR Breakthrough days	VaR Breakthrough rate%
ARMA	5	1.98
GARCH	5	1.98
POT	35	13.89

From the data in Table 5 and Table 6, it can be seen that the risk coverage rate of the parametric model has passed the backtracking test, but the POT model based on the extreme value theory has a very low coverage rate on VaR.

7 Result analysis

According to the results in Table 2 and Table 4, we can find that the S & P 500 Index in 2019 is in the yellow light area required by the Basel framework under the parametric model, and in the red light area under the POT model. It can be seen that the VaR value obtained by the model obviously underestimates the risk. There are three conjectures:

(1) The model is based on the ideal market conditions. For the rate of return, only its lag effect and random factors are considered. In reality, the financial industry is also affected by political circles, industry and commerce, and these influences are not quantifiable. Although they are all considered as random factors of the model, there is a mutual relationship between these factors. Therefore, they cannot be ignored and considered comprehensively.

(2) Since 2017, the international financial market has been in an unstable pattern. Although it has shown some vitality in some markets, the overall pattern of the financial market is still not optimistic. Most of the enterprises covered by the standard & Poor's 500 Index are large-scale enterprises in the United States. However, these enterprises have been affected by the financial crisis in 2020, and some enterprises have not extricated themselves from the crisis so far, resulting in the asset risks of enterprises still very large, and the model estimation is in line with the actual situation.

(3) Since the sample data used in the model is the daily trading closing price from January 3, 2013 to June 30, 2020, in which the abnormal fluctuation of financial market in 2020 is also used in the model calculation, it may have some impact on the estimation of the model.

Compared with the models established by the three methods, although the parametric model may be more able to reflect the risk situation of the market and has higher coverage for VaR, it is considered that the parametric model may not fully describe the real situation of the data due to the low R value obtained by the parametric model. But the POT model can fit the data well. We can guess that if we only study the risk situation of the market, we can adopt the parametric model, but if we study the tail situation of the whole market, the POT model may be more appropriate.

8 Follow-up discussion

As an indicator of the financial risk measurement, VaR also has some disadvantages:
First, VaR cannot measure the size of any potential loss that exceeds the VaR level. This will lead to the investment decision-making to optimize the expected profit in the limited situation when the VaR does not exceed a certain threshold, because the potential "rare case" loss exceeding VaR may be extreme.

Second, VaR is not a continuous variable, as is described by Artzner et al. (1999). This conclusion is due to the lack of the sub additivity of VaR.

Another common risk measure is Expected Shortfal (ES).

ES has obvious advantages over VaR: it does not involve aspects beyond the VaR level, and ES is a sub additive continuous variable. The attribute of the sub additivity means that the ES of the portfolio cannot exceed the sum of the sub portfolios of ES. The sum of these independent ES generates a conservative risk measure for the entire portfolio.

In addition, the incremental VaR, the marginal VaR and the component VaR can also be used to estimate and predict the financial risks.

References

1. Jorion P. Value at Risk VaR [M], Beijing: *CITIC Press*, 2005
2. B Hakwa, JA Zewicz, B Rudiger - Measuring and Analysing Marginal Systemic Risk Contribution using CoVaR: A Copula Approach [J], *Econometrica*, 2012, 50: 987-1007
3. Lennart F. Hoogerheide, Herman K. van Dijk. Bayesian Forecasting of Value at Risk and Expected Shortfall using Adaptive Importance Sampling [J], *Tinbergen Institute report*, 2008-092/4
4. Jonathan D. Cryer, Kung-Sik Chan. Time Series Analysis: With Applications in R [M], *Springer*, 2008
5. Li Yong, Ni Zhongxin. Bayesian test and model selection of financial arch model [J], *Chinese Journal of Management Science*, 2008, 16, (6): 25-28
6. Zhu Huiming, Li Feng, Yang Jinming. Analysis of Bayesian heavy tailed financial stochastic volatility model based on MCMC simulation [J], *Operations Research and Management Science*, 2007, 16, (4): 111-115
7. Shi Daoji. Practical extreme value statistical method [M], Tianjin: *Tianjin Science and Technology Press*, 2006
8. Gao Song, Li Lin, Shi Daoji. POT model of stationary series and its application in exchange rate Value at Risk [J], *Systems Engineering*, 2004, 22, (6): 49-53
9. Zhong Bo, Chen Ke. Using threshold model based on Bayese estimation to calculate VaR [J], *Journal of Beijing Technology and Business University (Nature Science Edition)*, 2008, 26, (5): 71-74
10. Zhao Yan, Li Hongwei, Peng Shijian. VaR Estimation Based on Bayesian MCMC Method [J], *Statistics & Decision*, 2010, 7: 26-27
11. Su Tao. Research on the improvement of VaR Measurement Method of financial market risks [D], Tianjin: *School of Management of Tianjin University*, 2007
12. Ouyang Zisheng, Mo Tingcheng. Research on the Internet financial risk measurement and evaluation [J], *Journal of Hunan University of Science and Technology (Social Science Edition)*, 2016, 16, (6): 25-28
13. Wang Chen. Research on the regional correlation and risk measurement of China's financial market volatility [D], Jilin: *Jilin University*, 2010
14. Wang Lifeng. Application of ARCH model in forecasting stock price fluctuation [J], *Statistics & Decision*, 2006, 4: 142
15. Meng Xiaohua. Modeling of heteroscedasticity model [D], Hubei: *Department of Probability Theory and Mathematical Statistics, Huazhong University of Science and Technology*, 2006

Design of Multimedia College English Teaching System Based on Virtual Technology

Qingxiang Duan

Zhengzhou University of Science and Technology, Henan, Zhengzhou, 450064, China

Abstract. With the development and application of virtual technology, we should integrate it with college English teaching to construct intelligent and three-dimensional English classroom and increase the interactivity of English learning, give students more vivid learning space. By designing a multimedia college English teaching system based on virtual technology, the English teaching pattern will be vividly upgraded. Based on the types of multimedia college English teaching system based on virtual technology, this paper puts forward the design scheme and application strategy of Multimedia College English teaching system.
Keywords: Virtual Technology, multimedia, College English, teaching system, design strategy.

The innovation of information technology has attracted the attention of all circles of society, and has also attracted the attention of scientific researchers. In China, information technology can be said to be widely used, not only greatly improving the level of teaching, but also improving the ability of students to learn independently due to the continuous innovation of information technology, virtual reality technology emerges as the times require and improve the level of science and technology. Especially in college English teaching, virtual reality technology plays an important role.

1 Types of multimedia college English teaching system based on virtual technology

The traditional English distance multimedia teaching system has poor interactive ability, which leads to students' low enthusiasm for English learning. Therefore, this paper designs an English distance multimedia teaching system based on virtual reality. The system uses 3D scanner to aid modeling by 3D inspection, and 3D visual display equipment to display virtual teaching 3D scene, the interactive devices such as gloves and joysticks send learning instructions and receive feedback of teaching information, so as to upgrade the intelligence of college English teaching.

1.1 VR + Micro lessons

Compared with the traditional micro-course, the micro-course based on VR technology is superior to the traditional micro-course. First, the vast majority of VR content is 3D, with 3D animation, a strong sense of immersion, a sense of reality for the learners to experience, it is helpful to solve the problem that the head-up effect of ordinary micro-lesson leads to the learner's lack of attention and easy to be distracted, which is also the great significance of VR to situational teaching. Secondly, compared with the traditional micro-lesson, the interaction of VR micro-lesson can arouse the learners' interest more effectively. For example, the experience of a VR course involves the interaction of knowledge points, which stimulates the learner to explore further and helps solve the problem of learning depth. In the process of teaching English, teachers can put the prepared VR content before or during the class, for example, in an English class that gives an overview of British and American culture, through VR to watch the content of the foreign life scene, feel the local folk custom. Although not abroad but can be demonstrated through VR panoramic video experience, classroom teaching to pave the way for a good cultural background.

1.2 VR + Flip the classroom

With the development of Internet technology and the rise of online classroom, the traditional classroom has been challenged. As the name implies, students learn to listen to class and take notes after class. At the same time, they put exercises, discussions and questions after class into the class, through teacher guidance, answer questions, interaction, overturned the order of the traditional classroom. VR lessons are also reversed, with students learning VR Lessons in advance according to their teachers' arrangement after class, and then spending more time face-to-face with teachers in class. Although there is not a lot of college English VR teaching content on the market, with the development of VR technology, there will be more and more courses for reference in the near future.

1.3 VR + BYOD teaching mode

A "BYOD"(Bring Your Own Device) is a Device that you use to learn on Your Own. These devices include a personal phone, computer, tablet, etc.. BYOD teaching model is a new concept, the main features: students can take their own devices anytime and anywhere learning; teachers and students in the device data and their own device data without interference. On the one hand, VR can be part of an E book, and you can put VR English content into a VR all in one box, like a VR glasses box, and you can plug your phone into that box, teachers can make it very easy for students to use VR lessons to improve their learning enjoyment. On the other hand, students will be able to access their teachers' online cloud systems directly via their mobile phones or ipads to view their teachers' pre-prepared VR lessons. English teachers can spare some time to experience VR English learning with students. Instead of just explaining grammar and vocabulary, they can "touch" knowledge with students.

2 Design of multimedia college English teaching system based on virtual technology

When constructing the multimedia network platform for English teaching, we must take into account the characteristics of English teaching itself. The characteristics of English teaching determine that more emphasis should be placed on interactivity in network-based teaching, but at present most of the network courses only have the function of presenting knowledge or content, such online courses are basically just changing the content of books in the medium of presentation, which is not fundamentally different from previous training methods such as broadcasting, reading extra-curricular books, memorizing words by oneself, and practicing oral English on one's own. In order to improve the quantity and quality of students' English skills training, we take full account of the interactivity between students and teaching contents, students and teachers, students and students, when we build online courses, and according to the following principles to investigate: (1) whether there is an effective incentive mechanism for learning; (2) whether there is an evaluation function for learning activities, whether this evaluation function is based on a sufficient and effective record of the student's learning activities; (3) whether the student is in control of the learning process; and (4) whether there is sufficient interaction between teaching and learning, in particular, to provide real-time effective feedback to students' learning activities; (5) whether can make full use of information technology and intelligent technology, develop high-quality open practice, improve students' learning interest and learning efficiency.

In the school teaching process, through many multimedia facilities has entered the language learning integration and the new process. The main features of the multimedia teaching model designed in this paper are the use of TV channels, watching online movies, using an electronic dictionary and configuring a web log in which students can upload homework or ask questions, to get feedback from peers and teachers. Video teaching. During the production process, make sure that each class is allocated an online movie time, ask students to watch a movie related to this class, and upload the movie to the multimedia cloud beforehand, because the instructor ensures that the video information can be played in class without wasting time. After watching a part or an entire episode of the movie, students were asked to share their reading ideas (via words or pictures) . By using this model, we can activate each other's memory, share opinions, and check and compare their understanding of visual text with that of written text, students use dictionaries as a reliable source of word translation because it allows them to make certain guesses about the meaning of new words. Thus, if guided in a controlled manner, the use of dictionaries can act as facilitators of the language learning process without making the learner overly dependent on them. Based on this assumption, students can use their dictionaries wisely after guessing the meaning of words. The use of the Internet to clarify unknown subjects in pictures also helps learners to master unrecognized English words and subjects and to share them with other students. When reading English text, students are asked to guess the meaning of words and sentences first, and if the definition is not clear or if they need to check their guess, they will use an electronic dictionary on their smartphone.

Ebooks. Learn from ebooks by setting up English blogs, sharing English e-books, and sharing material discussed in class or movie reviews with pictures and new words. In addition, some assignments are assigned to students through a web log. Students can use weblogs to ask questions and share ideas and experiences. Another important feature of weblogs is a list of useful websites introduced by teachers that students can use to improve their language skills.

Data collection and analysis. The core data of the system is collected by means of network log and information click, and the classified algorithm is used to process the data and provide data support for teachers to make the next step teaching plan, the core algorithm of the module is as follows: through the above system functions can change the way students learn foreign languages. The use of multimedia and the Internet has brought a new atmosphere to the classroom, which is more interesting to the students. The electronic dictionary is fast, easy to access, user-friendly, and provides students with more information, such as the source of the word, pronunciation/pronunciation, quick access to synonyms or opposite word meaning and definition; Video instruction brings real language to the classroom and provides students with input such as words, pictures, sounds and contextual languages; Internet access in the class gives students the opportunity to search for difficult content, and learn more about them by using the information they can find. E-books (weblogs) help students connect with other students and lecturers during and after class hours.

3 Application of multimedia college English teaching system based on virtual technology

In the process of college English teaching, curriculum resources are very important and directly related to the quality of teaching. In order to increase students' knowledge reserves and enrich teaching materials, English teachers should make good use of multimedia technology, collect and integrate English materials, and lay a good foundation for creating ecological curriculum environment. In practical teaching, teachers should analyze the curriculum resources in advance and decide the direction of developing extra-curricular English resources according to the key and difficult points in the textbooks. Then, through the Internet to find and collect relevant English materials, teaching materials and extra-curricular materials will be integrated together through multi-media tools to carry out teaching, so that students can more deeply understand the course content, it can also deepen students' understanding of English culture and improve the teaching effect.

In the process of college English teaching, in order to make the teaching environment ecological, teachers must change the traditional teaching method and make use of modern network multimedia technology to optimize and innovate. In the field of English teaching teachers can choose to flip the classroom teaching methods, in the organization of English activities, you can use multimedia tools to turn the course content into micro-class video, in this way, students can get a more intuitive understanding of the English curriculum content, and can study and explore the content independently, so as to give full play to their role as the main body. According to the video content of English micro-class, students can preview in advance, providing them with very convenient learning conditions, and can study and think on the course content independently. In addition, the teacher can also use the network multimedia, to broadcast English classics or movies to the students, which not only can enrich the English resources outside class, but also can let the students appreciate and analyze, so as to master more English expression skills, continue to deepen the understanding of the cultural connotation of English.

Classroom atmosphere is very important for language learning. A pleasant and lively atmosphere is more acceptable. In the classroom, teachers should strengthen the interaction with students, become students' friends, harmonious teacher-student relationship is easier to improve teaching results. Therefore, communication and communication in daily life are also very necessary. Teachers should give full play to their guiding role, promote the learning atmosphere in class, and infect everyone with their enthusiasm, to the student's outstanding performance must prompt encourage, between the teacher and the student harmonious relations, also has the very big influence to the student's study interest.

In the process of college English teaching, teaching evaluation plays a very important role. Teachers should use multimedia tools to optimize the evaluation system and strengthen innovation according to students' learning characteristics and needs, further promoting the ecological construction of English curriculum system. In the actual teaching, the teacher can use the multimedia, establishes the on-line appraisal system, carries on the on-line appraisal appraisal to the student's study situation. In order to optimize the teaching evaluation environment, teachers can record the whole teaching process of the course, which not only can understand students' learning performance in more detail, but also can play the recorded video to students during the evaluation activities, let the students see their own learning performance, so as to strengthen the thinking of cognition, self-analysis summary.

4 Conclusion

The teaching mode of college English course based on multi-media and network technology can improve students' learning interest, increase students' learning input, and promote students' independent and individualized learning, it is of great significance to solve the problem of less interaction between teachers and students. Since the use of multimedia technology to assist college English teaching is still in a rising and developing stage in the design, production and application of multimedia courseware, the sharing of multimedia teaching resources, the establishment of new teaching models and teaching evaluation, etc., it still needs to develop and improve itself.

References

1. Liu Lulu. The application of virtual reality technology in college English listening and speaking courses in application-oriented undergraduate colleges [J]. Electronic Testing, 2016, (23): 104+115.
2. Zhou Qing. Research on the application of virtual reality technology in college English teaching [J]. Caizhi, 2017, (28): 186.
3. Jin Yan; Yang Mingtuo. Designing a virtual learning space for college English based on the Virtools development platform [J]. Foreign Language Teaching, 2017, (4): 17-22.
4. Zhang Ning; Liu Yingchun; Shen Zhipeng; et al. A review of the application of virtual reality technology in the teaching of English for special purposes [J]. Computer Science, 2017, (S1): 43-47.
5. Zhou Bo. A brief analysis of the innovative path of higher vocational English immersion teaching method based on virtual reality technology [J]. Modern Communication, 2018, (19): 177+176.

The Value Embodiment and Realization of Art Education Function in Modern Society under the Background of Internet

Jianhui Liang

Guangzhou Academy of Fine Arts, Guangzhou, Guangdong, 510000

Abstract. With the development and application of "Internet +", intelligent technology has been widely and deeply integrated into various fields, and has become an important driving force for social development. Under the background of the new Internet, art education and communication have formed a new form of "online and offline". Under the modern social scene, in order to give full play to the function of art education, we need to take the new social scene as the foundation, and reflect the value of the function of art education in modern society, put forward the concrete realization path of the function of art education.

Keywords: Internet background; art education function; modern society; value embodiment; realization strategy.

Art itself is a special form of social history, and the creation of art is a more abstract spiritual creation, so art works have unique aesthetic value. Art as a humanistic method and method can attract people to be noble. Art production is to satisfy people's aesthetic needs by creating art works with aesthetic value. Art embodies the aesthetic consciousness of human beings, condenses people's aesthetic relations to the real world, uses the means of the Internet to give full play to the function of art, and carries out art education and art communication, is necessary for the development of society and the individual.

1 Under the background of the Internet, the function of art education in the value of modern society

The question which the internet produces needs us to use the internet thought to crack, needs us to carry on an internet time thought revolution. In a sense, the most important thing the internet has brought to our times is growth. The internet liberates all kinds of things from their fait accompli, fixity, isolation and temporality, and enables the essence and value of things to emerge and grow in the time and space of human beings as a whole, the creative potential of the human subject can be expanded and blossomed. And this growth is precisely in the interaction between different things and the boundaries of the expansion and completion. If we want to explore the issue of the artistic boundary in the Internet age, we should first firmly grasp this growth, starting with uncertainty, possibility and creativity, to dialectically understand the external boundary of art from positive and negative aspects. At present, the most important thing is to analyze and deal with the relationship between four aspects:

One is art and technology. The history of art shows that the change of tools, the renewal of materials and the progress of skills can give birth to a new art form with more vitality. Internet technology has brought about profound changes in the value chain of artistic creation, production, dissemination, appreciation and consumption. On the one hand, modern technology has given new wings to the development of art, making it fly higher and farther. Various forms of network art, digital art, virtual art, fusion art and other full load and display the human aesthetic ability, to show us the amazing and attractive aesthetic vision. On the other hand, the pace of artistic technicalization is also quickening obviously. Aesthetic Computing, neuroaesthetics, nano-art and biological art are pressing forward, which makes people feel that art has the possibility to sink into and disappear in technology.

Second, art and media. Art is a kind of spiritual presentation of the subject, and media is the spiritual communication between the subjects. Art not only has its own media attributes, and art only through the media to reach the audience. In the Internet age, the media fusion is accelerating, the media and the art two-way entry and the fusion, unceasingly to the deeper level evolution, the media art characteristic is more and more prominent. All this changes the traditional relationship and boundary between art and media directly or indirectly. In the framework of the Internet, media expansion, art media and media art constantly intensified, so that the independence of art, uniqueness, transcendence threatened.

Third, art and life. Art comes from life, and at the same time it is higher than life. However, the boundary between art and life has been broken in the pre-internet era. The Internet has greatly accelerated the pace of artistic life and artistic life, more and more artistic elements into daily life consumption, more and more life behavior and supplies directly into art. Art can not be separated from life, but it can not be lost in and out of daily life. For the tearing, sticking and mixing of art and life in the Internet age, we need not only to actively construct and stick to the value of art noumenon, but also to be tolerant and patient, strive to maintain a reasonable tension between life and art, to maintain a hierarchical distribution pattern of art.

Fourth, art and the public. The Internet has pushed and integrated the public's artistic demand, artistic creation and artistic consumption into a large platform of the human spirit and has been greatly satisfied, however, the elite's artistic creation, artistic expression, artistic communication and artistic consumption are being eroded, and the distance and boundary between art and the public are being drawn closer and rewritten. The old classic model of production has been transformed by the Internet. As the pace of art merging with the public and the market is accelerating, the influence of consumption-oriented, benefit-oriented and communication-oriented on the development of art is becoming more and more important.

2 The new changes of the function of art education in modern society under the background of Internet

In the process of traditional art communication based on "Internet +", we transfer the traditional art information to the outside through the network media and other new terminals. In this process, the traditional art communication will certainly have the virtualization characteristic to appear. According to the operation rule and the operation rule to carry on the substantial content and the specific code transformation process, has provided the very big space for the traditional art dissemination virtualization. Data becomes an important means of recording, transforming important information, such as words and pictures, into virtual data patterns.

From the perspective of external forms of expression, traditional art communication in the "Internet +" background has a variety of characteristics. The traditional forms of art communication have diversified characteristics, which to a certain extent promotes the rapid improvement of network information technology, with the rapid development of information technology, more and more people have gradually mastered the way of information dissemination and expression. On the basis of the rapid development of the network society, its content should be transmitted outwards through the audio-visual mode. In the Internet Society, the spread and diffusion of traditional art not only let the traditional text, animation and graphics, image and audio, video and other methods to spread out, and will be able to two or more traditional art forms of communication with each other, and finally reach a diversified interactive scientific and effective development of traditional art communication, traditional art communication to improve.

As a medium facing public space, media has become more and more dominant in the connection between ordinary people and traditional art. Since the technical means of communication have various influences on the appreciators, the media are increasingly expressing directly their evaluations of the objects of communication and making these evaluations enter the audience's horizon at the same time. In the face of the coming of the Internet Society, we should avoid superficial preaching and technocracy through comprehensive, systematic and comprehensive education, and continuously deepen the restriction and purification of bad culture, thus strengthens to the foreign culture selective absorption and to the national culture promotion. Through the way of interactive communication of information technology, the original demarcation of the subject of art education communication is blurred, and the way of interactive online makes the EDUCATEE both the acceptor of Network Art, is also the creator and the carrier of network art. Educational process, in short, is the formation of the ideological and moral process of the educated, compared with the teaching process with greater complexity.

As one kind of ideological and moral education environment, the network environment has the spontaneity and permeability to the influence of people's ideological and moral character. The tension between educational philosophy as a theoretical form and educational practice as a practical form provides impetus for the continuous innovation of educational practice. The network art is the human society development enters the high-tech Information Age, in the contemporary art occurs the digital revolution process to form. Network art education is a highly integrated and unified educational process of science and art. It is a process of contradictory development and change of knowledge, emotion, Will and behavior from maladjustment to adaptation, from imbalance to balance. Art is the concentrated embodiment of aesthetic emotion. In the process of art appreciation, psychological factors such as thinking, emotion, feeling, perception, representation, association and imagination are extremely active, therefore, art education plays an important role in the cultivation and improvement of people's emotion.

3 Under the Internet background art educates the function in the modern society the value realization path

With the emergence of digital art, artificial intelligence, experimental art and other art forms, the flexible, interactive and convenient information art forms have made the public more fond of the integration of art and technology, let the art and the public to find a perfect point of convergence, create a perfect visual experience. The popularization of art means to make certain things close to reality, life and people. With the development of our society, the current cultural point of view has gradually turned to the masses. Art, as an important form of cultural expression, promotes the public to feel and enjoy art. At present, the Internet is accepted by the general public as a way to perceive the world and society through the Internet. Therefore, if the artist's works, connotation, emotion and so on are presented to the public through the Internet, the public will be prevented from being limited by economy, place and time, through the free, quick way you can quickly find their favorite works, from the overall improvement of the public's aesthetic values.

With the development of the Internet, the new development mode of art has become an inevitable subject. Dividing art into "on-line" and "off-line" art has become the New Order of contemporary art. The off-line art mainly comes from the social place, has developed to a relative peak in today, and has the limitation of place, time and so on. Therefore, with the rise of the Internet, its original art form as well as its theoretical characteristics, characteristics, will make contemporary art self-adjustment. Constantly from the social, artistic, cultural and other aspects of the significance of the Internet, the value, the essence of the proposed new understanding. Art follows the pace of the times through the way of the Internet. Only by integrating art with the internet and creating multiple values from it can we go further and further along the rugged and long road, to the state, society, individuals play a certain positive role, so as to improve the national comprehensive cultural literacy, promote the development of social history, promote the all-round development of people.

The integration of disciplines in art education is a necessary way to realize the value of "truth", "beauty" and "goodness" in art education. The aim of art education is to improve the aesthetic and humanistic qualities of students, so more attention should be paid to its value rather than its knowledge and skills, and to the construction of an open, integrated, interactive and complementary art education system, and not just their own kind of art or some kind of art skills. In art education, the level of humanistic accomplishment inevitably leads to the difference of stand, idea and intelligence, and further determines the formation of artistic aesthetic experience with rational thought and strong emotion. At the same time, art education should not only obtain aesthetic experience through art aesthetic and art creation activities, but also reflect the thought and outstanding value transmitted by art works with the thought theory of humanities. The lack of integrity of art education results from the emphasis on art aesthetic practice and the lack of humanistic theory, or the lack of artistic aesthetic practice. Therefore, art education should pay attention to the infiltration and integration of art categories, aesthetics, pedagogy and other disciplines, only in the integration of disciplines in art education, in order to give full play to the aesthetic function of art education and realize the function of "truth-seeking", "beauty-worshipping" and "Goodness-seeking" in art education.

"discipline-based" art education model is to build a cross-disciplinary integration of the art education system. It is based on the integration of four humanities: Art Creation, art history, art criticism and aesthetics, in which art creation gives students a direct experience of empathy, integrating the imagination, sensitivity and creativity of art into the process and work of art creation; the history of art urges students to pay attention to the dimensions of the development and evolution of art history, art criticism aims to reveal the theoretical cognition and social function value of art aesthetics and creation activities; A philosophical inquiry into aesthetics guiding students to establish human life and emotional activities. It needs to be emphasized that whether it is aesthetic theory at the general level, or aesthetics of various art categories such as drama aesthetics, film aesthetics, etc. , therefore, it is necessary to adapt the aesthetic theory as a discipline to the aesthetic general knowledge as a humanistic general education, and to turn more from the metaphysical philosophical level to the combination level with the artistic practice.

4 Conclusion

Art is the unity of image, emotion, reason and life, and art education should be permeated with such organic elements as emotion, thought, aesthetics and reason. Under the background of "New Liberal Arts", art education should be "human" education rich in humanistic spirit and scientific spirit. Its value is not only aesthetic value, but also cognitive value, moral value and ideal value, in short, that is to let students have the true, beautiful, good cognitive ability and develop the ability to truly and fully realize the art education "true", "beautiful" and "Good" value.

5 Acknowledgements

1."Analysis and Realization of Modern Moral Education Value of Modern Lingnan Art Resources" from Key Scientific Research Projects and Characteristic Innovation Projects in Guangdong Province(2018WTSCX069);
2. "The Value and Application of Lingnan Culture in the Cultivation of Socialist Core Values in Schools in Guangdong" from the 2018 Education Science Planning Project of Guangdong Province(2018JKSJD057) ;
3.Special project for key fields of colleges and universities in Guangdong Province (Science and technology serving rural revitalization) "Theoretical and Practical Research on Art Intervention in Enabling Rural Revitalization in Art colleges and Universities"(2021ZDX4062)

References

1. Gaia Nan, Qian Fang. On the social service function of university art education [J]. Northern Music, 2019, (19): 250-251.
2. Huang Hai. Rethinking the discipline paradigm of art management [J]. Art Management (Chinese and English), 2019, (03): 22-26.
3. Deng Yiling. Diversification and function analysis of art forms in public space [J]. Western Leather, 2019, (16): 24-25.
4. Cai Yunling, Li Ailin. The artistic function and aesthetic value of film music [J]. Northern Lights, 2019, (4): 99-100.
5. Wang Lixin. How is Art Possible: The Cultural Practical Function and Aesthetic Form of Film and Television [J]. Journal of Chongqing Jiaotong University (Social Science Edition), 2020, (6): 76.

Application of Computer Network Multimedia Technology in Vocal Music Teaching Mode

Xizhi Liu

Shandong University of Arts, Shandong, 250014, China

Abstract. Nowadays, with the continuous development of the times, multimedia technology(MT) is rapidly infiltrating people's lives, and people are paying more and more attention to multimedia teaching, which has been well applied in vocal music(VM) teaching. Since VM is a highly practical and relatively abstract subject, it is necessary to organically combine theory with practice. In addition, nowadays people have higher and higher requirements for music aesthetics, and the enrollment scale of VM majors is increasing, making traditional teaching methods no longer able to meet the needs of vocal talents. Because the network MT can well solve the problems in the traditional classroom and make up for the shortcomings, it has played an important role in improving classroom efficiency, improving the quality of teaching and teaching effects. At the same time, the use of modern methods for VM teaching can also provide teachers with more convenience. Therefore, under this background, it is necessary to combine modern education methods with VM teaching mode(VMTM) to realize the transformation of VMTM. This article uses questionnaire survey method and data analysis method to make use of the advantages of computer network MT to make VM teaching simpler and more intuitive, so as to achieve better VM teaching effects. According to the survey results, most interviewees believe that the application of computer network MT in the VMTM has certain effects and brings a lot of vitality to the development of the course.
Keywords: Vocal music teaching, Teaching mode, Multimedia technology, Applied research.

1 Introduction

With the continuous development of society and the continuous advancement of science and technology, people have had a great influence on the mode of music teaching. Because the VM teaching itself is relatively abstract and complex, coupled with the simplification of teaching methods and the limited subjective comprehension abilities of both teaching parties, there are still some problems in traditional VM education methods. In the context of the new era, the combination of information technology and traditional VM teaching has become the trend of current social development. The rapid popularization of new media such as MT and online music has been favored by the masses, and the number of users has increased. It has also had a subtle impact on students' acceptance, information literacy, and interest in learning.

At present, the research results on the application of computer and MT in VM teaching are very abundant in the academic circles. For example, Liang Yuanyuan pointed out that the interactive mode of VM teaching based on network MT has achieved better practical results and improved classroom efficiency [1]. Yu Yanjie believes that the use of multimedia and network technology in VM teaching, combined with sound, video, animation and other elements, greatly enhances students' perception of abstract vocal theory knowledge and practice [2]. Han Linjiang proposed that the application of information technology to music teaching expands the depth and breadth of classroom teaching, and also increases the vividness and vividness of music classroom teaching [3]. Therefore, this article focuses on the application of computer network MT in the VMTM, and it has very important research significance and application value.

This article mainly discusses these aspects. First, it elaborates on the computer network MT. Then it discusses the VMTM and its current situation. In addition, the application of image noise reduction algorithm in multimedia VM teaching is also introduced. Finally, regarding the application of computer network MT in VMTM, a questionnaire survey and research were carried out, and the survey results and analysis conclusions were drawn.

2 Computer network multimedia technology and vocal music teaching mode

2.1 Computer network multimedia technology

Computer network MT, based on the computer network, combines the processing of sound and electrical information, multimedia images and other related content, and conducts comprehensive applications to integrate various rich and colorful resources into an organic whole, so that users can use multiple senses real-time information interaction with the computer.

With the rapid development of information technology, computer network MT has gradually been widely accepted by people and penetrated into all areas of life. At the same time, it is playing an increasingly important role in people's lives, and has gradually become one of the indispensable technical means for people's daily life and learning. Computer network MT has been applied in many fields, such as education, commercial advertising, and film and television entertainment [4-5], as shown in Figure 1.

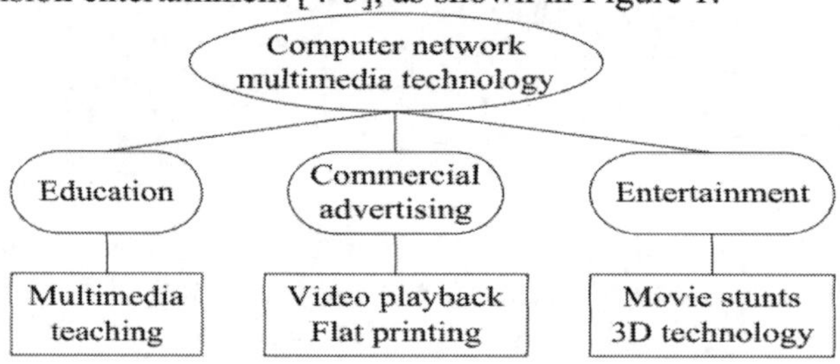

Fig. 1. Application of Computer Network MT.

Due to the emergence of network MT, the teaching mode has been changed, and students can use mobile phones, tablet computers and other devices for vocal training in the classroom. At the same time, it also provides teachers with a more convenient and comprehensive communication platform. In VM learning, students need to collect, organize and store a large amount of sound information. These audio data are processed through a computer, etc. and transmitted to the relevant system. In the classroom, multimedia equipment is used to conduct VM teaching through the combination of sound and picture or dynamic display [6-7].

2.2 Vocal music teaching

As the characteristics of each discipline are different, correspondingly, educational theories and educational laws are also different. VM also has the common characteristics of education, but its unique musical style is very obvious. While VM teaching is one of the teaching activities, it is also a practice activity of spiritual civilization education.

VM teaching also has these characteristics. The first is the social nature of practical activities. VM uses vocal singing as a means of expression to express music content, express the thoughts and feelings of lyrics and songwriters, express the singer's singing emotions, and reflect the current social situation at that time. Because of the combination of its rich melody and colorful lyrics, it is easier to be understood and accepted by people, and it can further form its wide-ranging social characteristics. The second is uniqueness. According to the different conditions and qualities of VM teaching objects, VM teachers adopt different training methods and training methods, and analyze specific problems in order to achieve corresponding teaching effects. The third is roundness. The teaching process of VM is very abstract and subjective. It requires the learners themselves to constantly, repetitively practice and consolidate until they master and develop the correct vocal habit. The fourth is cooperation. The mutual assistance, cooperation and communication between the two parties in VM teaching is an important way to achieve the best results in VM teaching. The importance of mutual assistance and communication between the two parties in VM teaching activities is far higher than that of other subjects. If one of them has a tendency to not cooperate, it is very difficult to achieve better teaching results [8-9].

In the teaching mode, the traditional VM class adopts the "word of mouth" teaching method, mainly through the description of teacher's speech or vocal demonstration, and students through their own understanding, continuous imitation and practice, and finally achieve the required vocal training state. This teaching mode is limited by the teacher's interpretation and self-expression ability and the student's comprehension ability, and the teaching method is abstract. If students cannot fully understand and adjust their own state correctly, it will easily affect the quality and effect of singing lessons.

Due to the particularity of VM teaching, the standard of the students' voice needs to be given feedback guidance in the first time. In the traditional VMTM, teachers and students train and guide vocal skills and vocal works in a fixed time period and space. Teachers can also make timely assessments of their own teaching effects based on students' on-site responses, follow up feedback information, and improve their own teaching methods.

Judging from the scale of education, in recent years, the number of art enrollment in colleges and universities has increased rapidly, and the quality of students seems to be uneven and uneven. As a result, some colleges and universities have transformed the traditional private VM lessons and traditional teaching methods, from traditional private lessons to group lessons and other teaching methods. Despite this adjustment, the number of teachers is still relatively small, so the per capita teaching time of VM classes is correspondingly reduced, the efficiency of face-to-face teaching also decreases, and the overall quality of VM classes is also declining [10-11].

The current situation of VM teaching is becoming more and more severe, and the problems of shorter per capita classroom learning time and relative decline in the quality of VM teaching have become increasingly prominent. VMTM will be restricted by time, scale, teaching method, teacher-student communication and other aspects, which causes the problem of general decline in teaching quality. Therefore, continuous improvement of VMTM and further improvement of the quality of VM teaching has become a very urgent issue.

2.3 Application of image noise reduction algorithm in multimedia vocal music teaching

The rapid development of network technology has led to the rapid growth of multimedia data on the Internet. Image has become the most basic and most extensive form in the process of visual information transmission. Compared with data types such as text and sound, digital images have higher characterization capabilities, and can store a large amount of color, structure, and texture information in a small amount of data. Among them, image noise reduction is a technical means to improve the quality of digital images, which is widely used in many aspects of computer technology applications, such as multimedia VM teaching. A good noise reduction algorithm can effectively improve the image quality. The image after noise reduction is smoother and the gray value changes more gently, making the teaching display screen clearer, thereby improving the quality of VM teaching and providing reliable technical guarantee for subsequent processing and analysis [12-13]. The image observation model is expressed as a formula.

$$W(u) = I(u) + M(u), u \in \Pi \tag{1}$$

Then, the variance of the image noise is.

$$Var\left(\frac{2}{|M_u|}\sum_{g\in M_u} M(g)\right) = \frac{2}{|M_u|^4}\sum_{g\in M_u} Var(M(g)) \tag{2}$$

The estimated value of pixel u is the weighted average of other pixels in the image, which can be expressed as.

$$MK(\hat{I}(\hat{u})) = \sum_{g\in 2} s(u,g)W(g) \tag{3}$$

Among them, M is an image contaminated by noise, I is a pure image, M is an independent and identically distributed Gaussian random variable, u and g are given pixels, s(u,g) is the weight, and M_u is the value of u The center of the field.

3 Questionnaire survey on the application of computer network multimedia technology to vocal music teaching mode

3.1 Questionnaire design process

The questionnaire survey selected 90 citizens of City F, including music majors, teachers and vocal lovers. Through the issuance of online questionnaires or paper questionnaires, the collection and quantitative analysis of the information filled in by users are carried out to draw conclusions of the questionnaire.

(1) In the preliminary preparation of the questionnaire, the number of questions should be as concise as possible to avoid fatigue of the interviewees.

(2) The questionnaire is released. Questionnaires were distributed through online questionnaires, on-site questionnaires, and inviting friends to help ask friends and students around them to fill in the questionnaires. A total of 90 questionnaires were distributed and 90 valid questionnaires were returned. The questionnaire recovery rate was 100%.

(3) Questionnaire analysis. Organize the collected questionnaire information to get the required information data. Analyze the results of the questionnaire. The analysis results include citizens' opinions on the application of computer network MT to VMTM. Some of the results obtained from the questionnaire are as follows.

3.2 Questionnaire survey content

The first part is the selected 90 citizens of City F, including music majors, teachers and vocal enthusiasts, to investigate the effect of computer network MT in the vocal teaching mode.

The second part is to sort out the information collected in the questionnaire and understand the opinions of 90 citizens on the application effect of computer network MT in VMTM. Part of the questionnaire survey results are as follows.

4 Questionnaire survey and analysis on the application of computer network multimedia technology to the teaching mode of vocal music

In this questionnaire, the interviewees included music majors, teachers, and VM enthusiasts, and conducted a questionnaire survey on the effect of computer network MT in VMTM. The survey results are shown in Table 1.

Table 1. Effect of computer network multimedia technology applied in VMTM.

Project	Very effective	Effective	General	Not effective
Teacher	11	10	7	2
Student	13	9	6	2
Music enthusiast	12	11	6	1

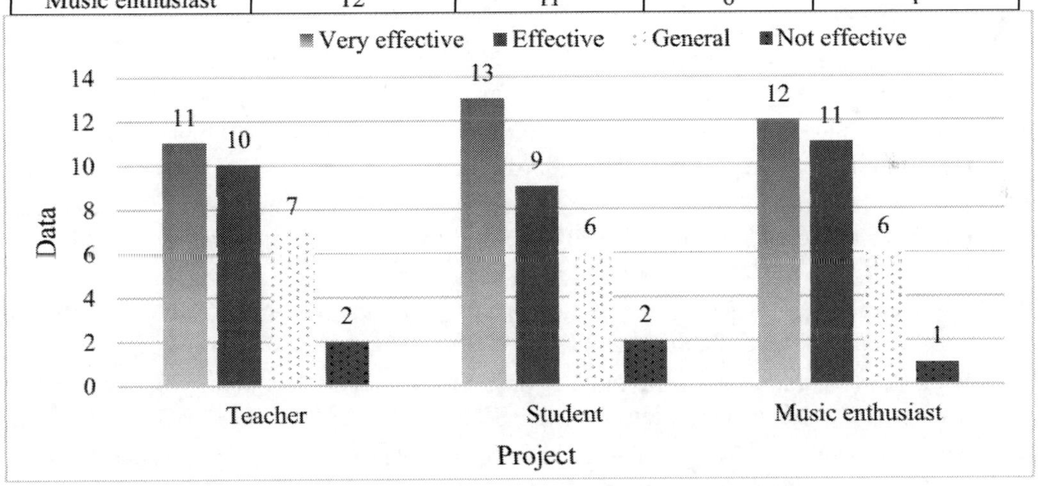

Fig.2. Effect of computer network multimedia technology applied in VMTM.

It can be seen from Figure 2 that with regard to the application of computer network MT in VMTM, 66 of the interviewees thought it was effective, while only 5 thought it was ineffective. Among them, a total of 22 students majoring in VM said that the application of the technology is effective. It can be seen that most of the interviewees believe that the application of computer network MT in the VMTM has a certain effect and brings a lot of vitality to the development of the course.

5 Conclusion

With the development and progress of the times and the continuous improvement of social demand, MT has gradually been widely used in the field of education, and has achieved good results. At the same time, MT also plays an important role in the spread of VM.

Computer network MT organically combines modern information technology and VMTM. In VM teaching, computers and other media devices are used for information processing and analysis, and then transmitted to students, providing students with more alternative learning platforms. At present, many colleges and universities have introduced computer multimedia teaching equipment to assist the development of school VM courses, provide an online communication platform for teachers and students, and help teachers improve teaching efficiency. Therefore, this article discusses computer network MT and VMTM, analyzes the application of this technology in VM teaching, so as to promote the innovation and improvement of teaching mode, and improve the quality of VM teaching and the efficiency of talent training.

References

1. Liang Yuanyuan. Research on Big Data and Contemporary Vocal Music Art in China: Discussion on Multimedia Interactive Mode in Vocal Music Teaching[J]. The Voice of Yellow River, 2019, No.538(13):69-70.
2. Yu Yanjie, Li Chengzhi. The application of multimedia and network technology in vocal music teaching in colleges and universities[J]. 2021(2011-1):74-75.
3. Han Linjiang. Application of Multimedia Information Technology in Vocal Music Teaching——Comment on "The Integration of Audio Music and Computer: Audio Music Technology" [J]. Chinese Science and Technology Paper, 2020, v.15(05):122-122 .
4. Dong Gaohong. The application of multimedia technology in middle school music appreciation class[J]. New Curriculum Guidance, 2019, 000(031): P.76-76.
5. Sun Yue. The application of computer network and multimedia technology in teaching [J]. A Guide to a Happy Life, 2017, 000(047): 89-89.
6. Liu Yu, Liu Rui, Jin Xin, et al. Application research of computer networking and multimedia technology in clinical teaching management[J]. Continuing Medical Education, 2018, 032(011):18-19.
7. Liu Ruirui. Research on the Application of Network Multimedia in Vocal Music Teaching[J]. Journal of Qilu Normal University, 2020, v.35; No.201(05):63-67.
8. Wang Yang. Strategies for the effective use of multimedia technology in vocal music teaching[J]. Tomorrow Fashion, 2017, 000(020): P.235-235.
9. Xie Qianqian. Analysis of the production of the western opera digital teaching resource library website-examples of the application of digital network technology in vocal music teaching[J]. 2021(2014-1):136-138.
10. Liu Yujie. Application and Innovation of Multimedia Technology in Vocal Music Teaching[J]. 2021(2015-18):197-197.
11. Qu Jing. Application of Internet and Network Multimedia in Vocal Music Teaching[J]. 2021(2016-13):111-111.
12. Wang Rong. Discussion on the Application of Multimedia Technology in the Teaching of Students' Vocal Music [J]. Satellite TV and Broadband Multimedia, 2020, No.510(05):239-240.
13. Lei Jijing. On the application of multimedia technology in vocal music teaching[J]. Curriculum Education Research, 2019(44):249-250.

Integration Practice of Modern Information Technology in English Teaching

Gang Li[1], *Wenyi* Li[2], *Bing* Li[3]

[1]China West Normal University, Nanchong, Sichuan 637002, China
[2]Sichuan Tianyi College, Deyang, Sichuan 618200, China
[3]Nanchong Vocational and Technical College, Nanchong, Sichuan 637100, China

Abstract. By integrating modern information technology into English teaching, we can improve the efficiency of English classroom teaching in an all-round way, and at the same time create a life scene which is easy for students to understand, to stimulate students' interest and enthusiasm to participate in English learning, and to help English teaching activities to achieve a comprehensive breakthrough. Therefore, we must realize the important position of modern information technology in English teaching, give full play to the advantage of information technology application, and promote English teaching to realize the intelligent reform. Based on the advantages of integration and application of modern information technology in English teaching, this paper systematically explores the practical approach of integration of modern information technology in English teaching.
Keywords: Modern Information Technology, English teaching, integration practice, application strategy.

Foreign Language Educational Technology is not a technical entity in the literal sense, but in the practice of foreign language teaching, it is used as a means and guided by the theories of pedagogy, psychology and linguistics, aiming at language teaching, through the creation, use and management of appropriate technical foreign language teaching processes and teaching resources, as well as the improvement of educational information technology literacy of foreign language teachers and students, to promote foreign language learning and improve foreign language learning performance of theoretical research and practical exploration activities.

1 The practical background of the integration of modern information technology in English teaching

Foreign language educational technology are the result of transdisciplinarity methodology. The integration of interdisciplinary knowledge in the information age can enable technology, theory and methods to form the TPACK framework of educational technology for foreign language education, it is to construct an interdisciplinary foreign language educational technology guided by the theories of linguistics, pedagogy and psychology, based on the teaching method and based on the educational information technology. This is the new theoretical connotation of foreign language teaching under the condition of information technology, the theoretical core of foreign language educational technology, and the driving force for the formation of foreign language educational technology, it is an important basis for the establishment and existence of the discipline. However, the theory and method of foreign language educational technology should be applied to the construction of information-based foreign language curriculum and teaching practice in order to reflect the theoretical correctness and technological advancement of foreign language educational technology as a new subject teaching theory.

Information technology-based English education is no longer just to establish English teaching website, no longer just to provide web-based learning resources, but to provide an intelligent online learning platform (also known as online learning space) . There is an important difference between the intelligent online learning platform and the general network resources: the online learning platform has strong interactivity, it can not only provide feedback information for learners, intelligent interaction with learners can also provide a space for teachers, students, administrators and researchers to communicate. From the technical point of view, the basic characteristics of education informatization are digitalization, networking, intelligentization and multimedia.

Although information technology has been widely used in English education, but from the current situation, the information-based English education should be at the stage of exploration. The depth of integration of English teaching and information technology needs to be improved. When people talk about information technology, they often mean computing and web-based technologies. In fact, the connotation of information technology is very broad, including information representation, acquisition, transmission, storage, processing, processing and utilization of technology. That is to say, the use of storage devices to store information is also an application of information technology, even the use of language and paper records and storage of information is also information technology.

In recent years, with the continuous updating and development of computer and information technology, the application forms of information technology in English teaching are more and more abundant, the scope is more and more wide, and the fusion depth is constantly strengthened. At present, the integration of information technology and English education is gradually increasing. However, from the current actual situation of English teaching, there are still many misunderstandings, the depth of integration is not enough. For example, some English learning websites use the advantages of information technology to provide students with a large number of practice questions, which are not much different from those in traditional paper resources, there's hardly any information technology involved. Some learning sites offer learning activities that are still limited to traditional learning methods such as knowledge presentation, memorization, and practice. Oral English training for the purpose of learning activities are still stuck in the repetition, imitation, follow-up and so on. The information technology content of these practices remains low. Some learning websites have developed an assignment publishing platform for English teachers, which allows teachers to easily post assignments for students and also to monitor students' completion of assignments. Although this way has a high content of information technology, but in reducing the burden of teachers, but also invisibly increase the burden of learning students.

2 The practical value of integration of modern information technology in English teaching

In the process of studying and practicing the deep integration of English teaching and information technology, we should accurately grasp the connotation of information technology, we should fully understand the role of computer, Internet, artificial intelligence, big data and other hardware and software in the process of promoting English teaching and learning, and give full consideration to the effectiveness of the application of information technology, to avoid formality, go through the motions, to avoid the application of some practices; to observe economic principles, to avoid the complexity of simple problems, to avoid the application of information technology to solve simple problems; To track the effectiveness of the application of information technology, to constantly optimize the way of deep integration of English teaching and information technology, and to strengthen the theoretical and practical research on the deep integration of English Teaching and information technology, study the theoretical model and practical operation process and method.

By giving full play to the advantages of modern information technology and creating life-like situations, the students can bring familiar things and scenes into classroom teaching to eliminate the students' unfamiliarity of foreign language knowledge. At the same time, students are more willing to use English to express their thoughts and feelings, so that students' imagination can be stimulated. Today's English teaching is no longer just teachers to impart knowledge to students, more students themselves to acquire knowledge, teachers only as a guide to teaching, rather than a leader. The acquisition of knowledge not only through teachers this way, but also need students to understand and experience. In the process of learning, students can make use of the beneficial learning resources brought by information technology to carry out autonomous learning and improve their learning efficiency and learning quality.

Multimedia can show some vivid pictures and necessary scenes, and express abstract contents in an image way. This not only can stimulate students' interest in learning, but also can stimulate students' potential in the invisible. Through the combination of picture and sound to deepen the students' impression, help students digest classroom knowledge, improve students' autonomous learning ability. With the popularization of the internet and the application of computer technology in the field of education, the flipped classroom teaching model has become simple and practical. Turning over the classroom teaching mode can free the teacher from the tedious teaching knowledge, have more time for the student to answer the question, and is beneficial to the teacher to carry on the personalized teaching and the student study ability raise. Before the class, the culture teacher asks the students to complete the learning tasks required by the culture course and watch the assigned online video course, TED Talk, international open course or college-made micro-class. In the classroom, the teacher organizes the students to carry out "brainstorming", which combines the cultural knowledge nodes organically. Through visual stimulation, the students can make clear the connections between the main points of literature, to facilitate learning and memory

The use of modern technology to create language situations, so that students can truly feel the culture of other countries. For example, when I teach the sounds of British English and American English in the classroom, it is difficult for students to remember and distinguish them. However, if some American and British movies are shown, students can easily appreciate the difference between British English and American English, it can also help students to have a real experience of foreign culture and living habits in movies or short films, and improve students' oral expression ability in a more natural language environment. Based on the theory of "constructivism" and "symbol interaction", and with modern foreign language educational technology as the leading factor, to achieve "cultural content integration", "in-class and out-of-class integration", "online-offline integration" and "in-class and out-of-class integration" teaching philosophy. To construct the humanities general course suitable for the comprehensive development of students, to combine the activities of internalized understanding and externalized expression of language, to improve the output ability of language, and to realize the comprehensive improvement of students humanities accomplishment.

3 The integration and practice of modern information technology in English teaching

Make use of the network information to carry on the development study. Teachers can assign tasks to students, asking them to collect information from different countries, and to present and act in small groups. This kind of teaching method changes from the traditional one-to-many mode to the independent study mode, the student uses the information technology to seek the material, carries on the Ponder and the summary according to the material which seeks, finally displays through the performance. This kind of teaching method can not only promote students to improve their autonomous learning ability, but also cultivate their team spirit. Language Learning and application is a multi-sensory experience, and different media transmit language information through different sensory channels. The essence of information, including the generation, transmission and reception of information, can not be separated from the "double code" system based on vision and hearing, "English audio-visual-speaking" organically combines the comprehensive functions of vision, hearing and oral expression, and embodies its own characteristics at different stages. It not only strengthens the function of classroom teaching, but also advocates the advantages of network training, not confined to the contents of books and time and space constraints.

Teachers can upload courseware and important knowledge points to the internet through some carriers. Students who do not fully understand the knowledge in class can also download the information provided by teachers through the Internet, to consolidate what they've learned. If students in the review process not easy to understand the knowledge points, you can consult students or teachers through information technology, but also directly on the network to find the corresponding solution process. In the marking time, teachers can use the computer marking, is conducive to saving time, improve work efficiency. The evaluation of Students' electronic work through computer operation can make English teaching mode more comprehensive and scientific and more acceptable to students. This way can arouse students' interest and arouse their enthusiasm for learning. On the basis of the audio-visual training, the audio-visual training is carried out. Teachers use the digital language laboratory, follow-up reading, imitation reading, man-machine dialogue, dictation and other forms of listening training. After class, teachers provide students with a wealth of listening resources on the internet and extensive listening training to guide them to complete various listening tasks, including selection, pairing, judgment, filling in the blanks, answering questions and dictation, etc. .

In the current English teaching activities, due to the lack of adequate language practice environment and the influence of examination-oriented education, many teachers pay more attention to the practice of grammar and sentence structure, but neglect the use of language skills. In this traditional teaching mode, teachers are not only tired of teaching, students have no interest in learning, classroom efficiency can not be effectively improved, students are more and more resistant to English subjects, learning burden is increasingly heavy. Multimedia technology is the core of modern information technology, multimedia courseware can realize the combination of picture and text, the effective combination of audio, video, text, pictures and other carriers, which is conducive to improving students' language skills, students can not only develop a sense of language, but also train their pronunciation and speed effectively by listening to and reading textbooks repeatedly and dubbing their characters.

Educational information is of great significance to the optimization of English teaching. It can not only save a great deal of teaching time, but also provide sufficient time and space for students to discuss and think independently, it strengthens the interaction between teachers, students and students, and helps students better master the knowledge they have learned. At the same time, it can also use information-based teaching tools to help students sum up and conclude knowledge points, deepen students' understanding and mastery of teaching knowledge. Therefore, teachers should stimulate students' learning interest and improve their English learning level according to the content of English teaching and the students' actual learning needs. The application of modern information technology in English teaching can help students to choose their own learning styles and develop their personalities. Students can learn in accordance with their own learning habits with the help of teachers, teachers in the learning process to guide students in the development of their personality, in the learning process can be innovative, to find their own way of learning, to serve their own learning, students in the independent innovation of learning at the same time teachers to guide the students to cultivate their innovative ability.

4 Conclusion

In the context of core literacy, modern information technology is deeply integrated with English classroom teaching, which promotes the reform of traditional teacher-centered teaching structure and teaching model. This change will be instructive and Operability to the teaching of English in secondary schools, as well as to the implementation of the liberal education, it is of great importance to cultivate students' English learning ability and practical application ability, which will provide ideas and approaches for solving the problems existing in English classroom teaching in the new curriculum reform.

5 Acknowledgements

China West Normal University 2022-2024 item 38 of the 2024—— innovation and practice of the cooperative talent training model for applied majors -- taking business English as an example.
Research on the translation of international business contracts in the context of "Belt and Road" by the China West Normal University, No. 17YC427

References

1. Zhu Wenxuan. The inevitable turn of college English teaching research —— based on the comparative analysis of English teaching research in recent ten years. Foreign language audio-visual teaching, 2019(6): 46-52.
2. Meng Yu, Chen Jianlin. A study on the dynamic evolution of foreign language learning in the Information Age: a visual analysis based on CiteSpace [J] . Foreign Language Teaching Theory and practice, 2019(4): 34-40.
3. Yuan Li. Construction of Ecological English classroom in the Internet + ERA [J]. Teaching and management, 2019(24): 105-107.
4. Zhang Xuemei. Research on the integration of modern educational information technology and college English writing teaching —— comment on college English writing teaching and research based on digital platform [J]. Chinese scientific papers, 2020(9): 1118.
5. Liu Tingting, Lu Da. A study of innovative approaches to informationalized mobile learning in college English teaching from the perspective of information ecology. Information Science, 2020(12): 134-139.

Research on the Application of Ant Colony Algorithm in University Teaching Management Service System

Wei Yi

Zhengzhou University of Science and Technology, Henan, Zhengzhou 450064, China

Abstract. The ant colony algorithm is based on the real ant colony foraging behavior research algorithm, based on the ant colony algorithm university teaching management service system, through the development of one-click automatic course scheduling, drag-and-drop Manual course scheduling, the multi-mode, service-oriented perspective to promote college teaching management, greatly improve the efficiency of college teaching management services, fully adapt to the intelligent college teaching management environment. Based on the ant colony algorithm, this paper analyzes the background of the application of ant colony algorithm in university teaching management service system, and puts forward the application strategy of ant colony algorithm in University Teaching Management Service System.
Keywords: Ant Colony algorithm; University; Teaching Management Service System; Application Research

Biologists have found that the behavior of individual ants is interacting with each other. In the process of ants' movement, they can leave a substance called pheromone on their path, which is the carrier of information transmission and communication between individual ants. The integrated educational administration platform, based on the micro-service structure and centered on "service", provides educational administration management, educational administration evaluation, experimental and practical teaching management, and other educational services for colleges and universities, to meet the management needs of school year system, Credit System and school year/credit mixed system, to help students individualize the training of one-person-one-case, and to carefully manage the teaching "high quality".

1 Application background of ant colony algorithm in University Teaching Management Service System

Many scholars have done a lot of work to improve the performance of ant colony algorithm. Dorigo et al. then proposed the ant colony system (ACS). The rule of state transition of ant selecting city is improved, the pheromone is released after traversing only the ant that finds the optimal path at present, and the pheromone concentration on the path is reduced in the course of transition, and the premature convergence to the same path is avoided. MMAS algorithm, the pheromone concentration on the path is limited to [t min. In the range of

TMAX], the initial value of pheromone on each path is set to T Max, and P is smaller, so more feasible solutions can be found in the initial stage. This algorithm is one of the best models for solving discrete optimization problems such as TSP. The hybrid ant colony algorithm proposed by Gambardella et al, and the fast ant system proposed by Taillard et al all improve the optimization ability of ant colony algorithm under certain space complexity as far as possible, and widen the application fields of ant colony algorithm.

In the implementation of ant colony algorithm. The value of ραβ and other parameters determines the balance between the searching speed and the convergence speed. Dorigo et al obtained the best parameter setting of AS algorithm when $a = 1$, $\beta = 5$, $\rho = 0.5$, while in ACS algorithm $\alpha = 0.9$, $\beta = 2$, $\rho = 0.1$. Jiang Lingyan et al analyzed the effect of αβρ on the performance of the algorithm, and pointed out that the algorithm has better performance when $\alpha \in [0.1,0.3]$, $\beta \in [3,6]$, $\rho \in [0.1,0.3]$.

Guttjahr et al firstly proved the convergence of graph-based ant colony algorithm (Gbas) and improved GBAS/tdev and GBAS/TDLB algorithms. Stuezle and Dorigo proved that a kind of ant colony algorithm when the number of iterations tends to infinity, the algorithm can guarantee to find the global optimal solution. The convergence rate of ant colony algorithm is analyzed by Huang Han et al and ACS. Sun Tao and others have made a preliminary study on the convergence of a class of simple ant colony algorithm. Ding Jianli and others use genetic algorithm to generate better initial distribution of pheromones, and then use ant algorithm to find the exact solution. In this way, the cooperation effect of ant colony algorithm and the evolution effect of genetic algorithm can complement each other, and win-win both in time and optimization.

It is characterized by distributed computing and positive feedback mechanism. The distributed computing enables multiple computers to perform simultaneous computations, increasing the speed of the solution, and the positive feedback mechanism enhances the role of pheromones in searching for optimal solutions faster and saving time. Strong robustness, just a little change to the model, can be optimized for other types of problems. In order to improve the performance of the algorithm, it is easier to use in combination with other heuristic. Information can be exchanged between individuals, and the communication overhead increases less. Although the ant colony algorithm has been applied to many practical problems and has many advantages, it also has some disadvantages: the long initial search time of the algorithm affects the search efficiency. The pheromone of each path is the same at the initial time. Only with the increase of time and the positive feedback of information, can the difference of each path be reflected and the optimal solution be obtained. But this process usually takes a long time, especially in solving large-scale optimization problems, which takes a long time and affects the efficiency.

2 Functions of teaching management service system in colleges and universities

2.1 Student status management

The school educational administration office and the second-level teaching management unit can track students' comprehensive information during the school term through the student status management function of the educational administration management information system, including the basic personal information at the time of admission, retirement, promotion, transfer of Major, graduation and other changes in status information, and the above information for daily maintenance.

2.2 Management of lesson plans and lesson scheduling

The teaching plan is the teaching basis which the university teaching management department extracts according to the talented person training plan, the teaching management information system can realize the functions of inputting, querying, revising and printing the teaching plans of different grades and specialities, the system can automatically check each class, teacher and classroom conflict, the correct arrangement of each semester curriculum.

2.3 Management of online course selection

The educational administration management information system provides an equal and competitive platform for college students to select courses online. The educational administration office and the secondary colleges and departments set up elective courses, elective time and elective subjects in the system according to the teaching plan, in the network through the pre-election, by-election, withdrawal of the three stages, the completion of the campus network all students elective activities.

2.4 Examination and performance management

Examination is an important part of teaching management, the system can carry out the examination schedule, examination room and teacher invigilation arrangements, and automatically sum up the examination schedule, saving paper work. After the test, the system can input, revise, sum up and analyze the students'scores, and then the report forms and the list of students who need to make up or retake the test are derived according to the parameters of the average score and the highest score of students or classes.

2.5 Management of teaching quality evaluation

On the one hand, teaching administrators can systematically derive reports on students' excellent and passing rates in different subjects to analyze the quality of students' learning. On the other hand, students in the system of the message board for the school's teaching link feedback, and through the system to participate in the teaching management department organized evaluation activities to evaluate the teaching quality of teachers.

3 Design of Teaching Management Service System in colleges and universities

3.1 System network architecture design

This system is based on the Internet/Intranet Network Operation Mode, its main system functions are the network server, the front-end client uses the Web browser, the back-end server is the system service and the management core, it is often referred to as the thin client/fat server model. The functions of the server include: the realization of the functions of the system (Business Logic) , the Web services and application services to the client, the interface of the database system, etc. . Based on the micro-service technology architecture + flexible container expansion mode, a caching mechanism is introduced to ensure flexible expansion of resources, stable and efficient availability of the system, set up a fault query window, and quickly check out possible course selection failures, to achieve non-caton course selection.

3.2 System architecture design

From a developer's point of view, there is a tendency to use the b/s pattern for everything because it has a distinct advantage over other patterns. First of all, the use of standard browser in the client, no need to install client applications, can achieve zero installation, zero training, and cross-system, cross-platform capabilities. However, the content of the educational administration management system in colleges and universities is very complex, involving many aspects, managing a wide range, many modules are interactive and deal with a large amount of data, if all the B/s single mode is applied, user-friendly operation and use is difficult to solve, such as course scheduling management, its security is not as good as C/S structure. Therefore, the system architecture mainly considers three-tier C/s mode and B/s mode. The educational administration personnel's management interface uses the three-tier C/s structure, but the teacher, the student, the information inquiry, the information release and so on practical aspect is broad, the security and the interactivity is not high module uses the B/S structure, so we can take full advantage of each other.

3.3 System functional module design

The system is composed of 16 subsystems: resource management subsystem, Student Basic Information Management Subsystem, Course Plan Management Subsystem, Course Arrangement Management Subsystem, Course Selection Management Subsystem, student course selection subsystem, classroom result entry subsystem, Result Management Subsystem, examination management subsystem, student status processing subsystem, Experiment Management Subsystem, graduation and Degree Qualification Examination Subsystem, User Management Subsystem, General Inquiry Subsystem, English Test level 4-6 and computer level examination subsystem, FEE registration management subsystem.

3.4 Database design

The design of the database is mainly based on the system requirements and functional structure, in order to save all kinds of data used in the university educational administration management system, the data table in the database of university educational administration system can include the following categories: students, teachers, courses, administrators, grades, etc. . The data table and field should be defined according to the frequency of information updating and using and the need of programming. In the database to create the user registration data table, student data table, teacher data table, curriculum information table, performance data table and other data tables. Considering the extendibility of the system, the database can be used as a shared database for other teaching management, and the data tables can be associated with the same name fields.

4 The application strategy of ant colony algorithm in University Teaching Management Service System

The teaching management post personnel is related to the teaching operation of the department and even the whole school. It is an important guarantee to realize the informationization construction of colleges and universities whether they can skillfully use the educational administration management information system to serve the teaching management. Therefore, colleges and universities should establish an effective training mechanism, strengthen the computer training for teaching administrators, and carry out the teaching administration system training for the new administrators, raise the level and technical level of teaching management personnel. There are many functional departments, administrators, teachers and students involved in teaching management in colleges and universities. In the use of educational administration system, a modern means, the university administrators should first give attention and support, and promote the use of the system departments and teachers and students enthusiasm.

When developing and introducing the software of educational administration management information system, we should not only consider the comprehensiveness of the service for teaching and scientific research, but also consider the actual law of teaching management in colleges and universities. At the same time, timely collection of feedback from teaching administrators, teachers and students, and continuous modification and improvement of system functions and modules by technical personnel, to create a school to meet the actual needs of the school, in line with the characteristics of the educational administration management information system, truly achieve scientific system configuration. After the system is fully understood and put into use, colleges and universities should strengthen the implementation of the system to solve the problem that some functions are not fully effective, so as not to use manual operation when the system can be used. For example, in the field of practice and teacher management, the selection of topics and grades of graduation theses, the application and approval of teachers' transfer of courses should be managed through the educational administration system, so as to reduce the workload of teaching administrators, maximize the management efficiency of the educational administration system.

Comments on the PC side of teaching management: suitable for academic administrators. The multi-dimensional view visualization data analysis chart and the teaching quality appraisal data report, helps the appraisal teaching work to carry out in an orderly way, realizes the appraisal teaching process high-efficiency management monitoring. Evaluation and Teaching Management Mobile (small program): for students, student information, teachers, supervisors and other multi-role users. Free Evaluation, task evaluation, scanning code evaluation, classroom evaluation, evaluation process flexible and easy to use, real-time evaluation of teaching, improve the quality of teaching evaluation data. The evaluation management based on process evaluation, relying on multi-dimension evaluation index, multi-staff mutual evaluation and multi-times cumulative evaluation, provides reliable data support for improving teaching quality. It can not only control the quality of teaching process in real time, but also feed back the evaluation data to the intelligent analysis platform of teaching big data.

5 Conclusion

The application of ant colony algorithm in University Teaching Management Service System is easy to converge to the local optimal solution. This is mainly because positive feedback accelerates the deposition of pheromones. When the global optimal solution has not been found, all individuals may search for a completely consistent solution, but this is only a local optimal solution, so that the algorithm can not be further search. Although the ant colony algorithm has such defects, but as long as we find the reasons can be solved, such as appropriate addition and subtraction of pheromone positive feedback effects, and other algorithms combined. Through the analysis of ant colony algorithm, this paper provides a good foundation for the later improvement and application of teaching management service in colleges and universities.

References

1. Yang Yiyu. Design and implementation of college teaching management service system based on WeChat public platform development model [J] Information and Computer (Theoretical Edition), 2016, (7): 85-86.
2. Xiang Kaiquan, Chen Zhibo, Wang Minquan. Personalized teaching management service system and realization of big data [J]. Electronic Technology and Software Engineering, 2016, (7): 171
3. Liu Dejian, Du Jing, Jiang Nan, et al. The development trend of artificial intelligence integrating into school education [J]. Open Education Research, 2019, (5): 26-32.
4. Ai Juan. Design and implementation of college teaching information service system based on mobile Internet technology [J]. Computer Knowledge and Technology, 2018, (4): 33-42.
5. Lu Li. The technical realization of using WeChat platform to develop mobile teaching service system [J]. Information and Communication, 2019, (1): 285-286.

Research on the Development of Art and Rural Culture Revitalization from the Perspective of New Media Communication

Yong Fang, *Jianhui* Liang

Guangzhou Academy of Fine Arts, Guangzhou, Guangdong, 510000

Abstract. New Media is the witness of rural culture revitalization, and also an important force to promote rural culture revitalization. With the development and application of new media becoming more and more mature, it will be on the basis of enriching rural cultural life in an all-round way by giving full play to the advantages of new media carrier and communication, and promoting the deep integration of the revitalization of art and rural culture, keep the vitality of artistic creation, and consolidate the ideological foundation of rural revitalization. Based on the study of new media communication, this paper systematically discusses the development value and implementation strategy of art and rural culture revitalization and integration.
Keywords: New Media Communication; Art; rural culture revitalization; integrated development.

To cultivate civilized local customs, good family customs and simple folk customs, and to promote the revitalization of rural culture, is the meaning of the prosperity and development of rural culture. The local culture has a long history, is an important part of Chinese excellent traditional culture, is also generations of ancestors production, life and aesthetic concentrated reflection. Literary and artistic workers should deeply understand the rich connotation of local culture and create more excellent works with both literary quality and beauty in the process of rural revitalization.

1 The development value of the combination of art and rural culture revitalization from the perspective of new media communication

In order to revitalize the countryside, culture must go first. Rural culture revitalization is an important part of rural revitalization, which is not only the adherence and inheritance of agricultural culture and local culture, but also the innovative development of itself, thus creating a modern rural culture. The revitalization of rural culture is a soul-casting project, which is conducive to the realization of rural civilization, the construction of rural spiritual civilization and the satisfaction of farmers' needs for a better life, the revitalization of rural culture can promote the development of rural economy, politics, culture, society and Ecology Better, faster and more efficiently. Propagandize the policy idea, condenses the social thought common understanding. Propaganda work is an extremely important work of the party, and it is a fine tradition and political advantage for the generations of Chinese leadership to constantly seize the victory of revolution, construction and reform. Compared with other means of communication, network communication is not affected by time and space, communication platform is more convenient, discourse is more vivid, and it has a wider audience, it is an important channel for disseminating policy ideas to farmers, guiding ideological and public opinion, listening to the voice of the common people, reflecting social problems, answering questions and solving puzzles, and building consensus, it is related to whether the rural revitalization strategy can be accepted and supported by the broad masses of farmers, and whether it can be effectively landed in the process of implementation.

Broadening the main body of communication and promoting the status of farmers as the main body. The revitalization of rural culture is a complex systems engineering, with rural areas as the main battleground and farmers as the main implementers. For a long time, the phenomenon of rural aphasia and farmers losing their voices has caused the society to have a rigid understanding of rural culture and the image of farmers. In recent years, internet communication, represented by short video and live broadcast platforms, has attracted a large number of farmers to actively participate in order to make it more convenient, interesting, interactive and connected, a group of farmers, represented by "Li Ziqi," a young entrepreneur returning home, "Qiao Fu, nine sisters," a peasant woman, and "seven fairies of the romantic Dong family," created by Wu Yusheng, the first secretary of Gebao village in Liping County, Guizhou Province, have gone viral on short video platforms, local customs, cultural tourism and the sale of agricultural products have been brought to the fore, encouraging more farmers to take an active part in the revitalization of rural culture.

Innovating the content of communication and reconstructing the rural cultural ecology. Under the background of "technology empowerment" of Internet, the content mode of production of internet communication has also undergone profound changes, it constructs a new rural cultural ecological circle which is created by farmers as content creators and farmers who gather in the network platform. Rural excellent traditional culture, local accent and folk customs, beautiful scenery, delicious food and accommodation and other earthy topics become the new rural cultural material, and the simple and natural original ecological style becomes the new expression, it arouses the masses' spiritual yearning for the rural life, releases the farmers' internal demand of expressing the right of discourse, and reconstructs the traditional rural cultural ecology.

Eliminate the information barrier and improve the environment of information dissemination. Internet-based self-media dissemination, represented by micro-blogs, wechat, short videos and live broadcasts, has changed the previous dominant methods such as television, newspapers, bulletin boards and village cadres' propaganda, a huge amount of relevant information is broadcast in all-weather and all-media, breaking the "Information Island" effect of poor information and communication in rural areas, and realizing unlimited information supply and precise push, the cultural estrangement brought about by the dual structure of urban and rural areas has been removed. At the same time, the network platform gives farmers more voice and technical, flow support, help farmers conserve wisdom, improve the ability of media expression and flow cash.

2 The requirement of the development of art and rural culture in the view of new media communication

Pay attention to the inheritance and innovation of rural culture, improve the endogenous impetus of the revitalization of rural culture. To dig deeply into the ideas, humanistic spirit and moral norms contained in the excellent traditional culture of the countryside, and to find the contents that are compatible with the contemporary rural culture and in harmony with the modern society, strengthen creative transformation and innovative development to create a brand of rural culture. On the basis of protection and inheritance, we should accelerate the revitalization of rural cultural resources, find the correct orientation of characteristics, carefully plan the theme, create rural internet celebrities IP through financial support and flow support, enhance commercial value, and broaden the industrial chain.

Set up the consciousness of spreading fine works, excavate the development potential of rural culture revitalization. In today's era, the internet information is like a torrent boat, under the waves of sand, high-quality, high-value information can stand the dual test of traffic and content. To strengthen the sense of quality products, adhere to the "content is king", pay attention to user needs and user feedback, to provide users with more personalized, higher-quality content services. We will implement the project to improve the quality of internet-based works, strengthen the creation and production of rural cultural themes, and constantly introduce works that reflect the new image of farmers and show the new features of rural areas, so as to create works with height, depth and temperature, earthing gas, increase bottom gas, gather the high-quality work of popular.

Improving farmers' new media literacy and stimulating rural culture to vitalize talents. In the past year, Douyin's rural videos have garnered 12.9 billion likes, and the income of rural video creators has increased 15 fold over the same period last year, according to the first report on rural areas, released by 2021 on June 22. However, the report shows that among the creators of rural, rural and rural douyin, the proportion of urban white-collar entrepreneurs returning to their hometowns is the highest, and the role of farmers as the main body has not been fully played. To give full play to the role of the farmers as the main body, it is necessary to strengthen the construction of the farmers' new media network communication team, and focus on improving the professional qualities of the three farmers' anchormen in such aspects as ideology, morality, legal common sense and the ability to preside, we will strengthen cooperation with high-quality MCN institutions and internet celebrities to create high-quality cultural products in rural areas. We should strengthen the training of farmers' network information skills, enhance their ability to obtain, discriminate and control information on the Internet, and help them actively participate in the practice of rural cultural revitalization through the new media on the Internet.

As the main body of rural cultural construction, farmers should take the initiative to understand and define the social basis, characteristics and future trend of their own culture. In the process of cognition, judgment and reflection on their own culture, they dialectically look at different cultures and know clearly the characteristics, advantages and disadvantages of rural culture based on farming civilization. Through the cultural consciousness, the peasants can realize the profound sense of the rural cultural history, the emotional temperature sense, the humanistic feeling, arouse the deep cultural emotion in memory, and find the way to the future development of the rural culture. Arousing cultural identity is the starting point of remolding cultural values. The process of the reconstruction of rural culture is the recognition of "the life value with self-restraint, the life value close to nature and countryside, the production value of supporting green development, the ecological value of maintaining the harmony of nature, the educational value of optimizing personality, the humanistic value of sublimating homesickness and the social value of promoting public order and good customs".

3 The development strategy of art and rural culture revitalization and integration in the view of new media communication

In order to promote the revitalization of rural culture in the new era, we must combine the supply of public culture with the inheritance of local culture, excavate the resources of rural art and culture, and promote its creative transformation and innovative development, to promote the beauty of the vast countryside from the beauty of the moment to the long-term beauty, from the external beauty to the internal beauty, from the beauty of style to fashion, from the beauty of the environment to the beauty of development.

First, we should deeply excavate the resources of rural art and culture, and enrich the art carrier. The revitalization of art-enabled rural culture is not simply the preservation and renovation of one or more ancient buildings, the development of an art festival, or the implementation of a traditional agricultural cultural and creative project, or an overly simplistic cultural transplant. Many villages have accumulated the achievements of Chinese agricultural civilization, with distinct cultural personality and artistic characteristics. To empower rural culture with art, we need to sort out the context and mechanism of different rural cultures, in which both material and non-material are integrated, there are also aesthetic and practical, traditional and modern, man-made and natural integration, need to dig and clarify the rural art resources, so as to consolidate the original culture, wake up and activate the rural culture. With the help of modern information technology and innovative means, we can further enrich the artistic carrier, combine the construction of rural civilization with the inheritance of rural traditional culture, and develop colorful, positive and healthy rural art activities, highlight the performance of rural traditional culture, experience and appreciation.

Second, we should promote the deep integration of rural art and culture and tourism, and cultivate new cultural forms. According to the difference of cultural resources in rural areas, this paper seeks the artistic elements that are highly "cohesive" between rural characteristics and cultural creativity, develops cultural creative products and services with rural regional characteristics, and builds cultural and creative works. The rich historical and cultural resources, folk cultural resources and ecological cultural resources in rural areas should be artistically transformed to beautify the characteristic rural architecture and public art space, the combination of culture and art with rural production and life, the construction of cultural scenes, the creation of rural cultural brands, and the manifestation of "one village, one product" artistic features. Innovative Ways of integrating literature and tourism, fostering and developing new forms of art and culture, and promoting the classicalization and ritualization of folk cultural experiences, festivals and performances, etc. , and it will be integrated into the rural tourism areas, residential characteristics, rural complex and other projects in the construction. Using the new media communication platform to expand the marketing channels of art and Culture Tourism, turning the countryside into a tourist attraction and a tourist destination, and promoting the development of rural cultural tourism industry.

Third, we should promote the construction of rural art talents and take the rural art talent resources as an important support for the rural revitalization of the talent team. The revitalization of rural culture with artistic empowerment should not only be on the level of material culture construction, such as the establishment of rural museums, the organization of cultural performances and so on, but also strengthen the cultivation of rural art talents. We should further create an atmosphere to attract more talented people to participate in the cause of rural revitalization, and promote the development of rural culture and the arts and cultural industry. Through carrying out various activities to deepen farmers' understanding of the human geography, historical culture, art aesthetics and other aspects of the region, and to enhance farmers' cultural and artistic attainments as well as their sense of identity and self-confidence in local culture, promote their conscious participation in the cultural and artistic construction of the countryside. We will strengthen the training of artistic talents in the countryside, organize some old artisans, craftsmen and other folk artists to pass on their skills, and encourage more young people to take root in the deep countryside, lock in the lingering nostalgia and pass on their deep feelings of hometown, to pass on the old country culture and art.

Fourth, we should build a long-term mechanism for the rejuvenation of rural culture with artistic empowerment, and form a pattern of co-construction, co-governance and sharing. Culture and art endow the rural society with order, soul and vitality, which is the important motive force and source of igniting rural vitality and the important content and way of perfecting rural governance. We should constantly improve the talent, funds and rural art education and other related systems and measures to really produce art-enabled rural cultural revitalization of the synergy effect and system effectiveness. We should integrate the demands of socialist core values into the whole process of rural spiritual civilization construction and cultural and artistic production, so as to raise the level of rural civilization. Attaching importance to the construction of the mechanism for the establishment of rural cultural and artistic brands, and striving to build individual artistic individuals into a rural cultural complex and a spiritual paradise shared by the peasants, so as to meet the new expectations of the peasants for a better life to the greatest extent possible, let the broad masses of farmers live in the fields of hope full of artistic flavor.

4 Conclusion

The cultural revitalization is the key support for the implementation of the rural revitalization strategy, and the important source of the rural revitalization strategy. In the new era, all parts of China should take safeguarding the fundamental interests of the villagers and promoting their common prosperity as the starting point and foothold, and implement the strategy of revitalizing rural culture according to local conditions, focusing on the foundation of material culture, the soul of spiritual culture, the root of forced system culture, the root of behavior culture, the construction of a four-in-one approach to the revitalization of rural culture. The involvement of art in rural revitalization should fully combine the local cultural background, historical evolution, folk customs, residents' feelings and other factors, carry out the construction and transformation of Chinese culture and history, and highlight the characteristics of folk customs. As an important form of building a beautiful countryside, art is involved in rural revitalization as a long-term experiment, which can continuously put forward a rural construction plan limited by planning and design, waiting for the magical chemical reaction between the two.

5 Acknowledgements

1. 2019 "research on the innovative development of patriotism education in the guangdong-hong Kong-macao Bay Area in the New Era"(No. 2019 WQNCX054), a project of young innovative talents in universities in Guangdong province.
2. 2021ZDX4062"the theory and practice research of art colleges' involvement in rural revitalization"(2021ZDX4062)

References

1. Zhang Xinjiang. Research on countermeasures and effective paths of culture and art to help rural revitalization [J]. Music Creation, 2018 (12): 148-150.
2. Liu Zhigang; Chen Anguo. The conflict, fusion and mutual feeding of urban and rural cultures from the perspective of rural revitalization [J]. Administrative Management Reform, 2019 (12): 60-65.

3. Li Ming; Chen Qisheng; Zhang Jun. Path Construction of "Four-in-One" Rural Culture Revitalization [J]. Hunan Social Sciences, 2019 (06): 147-156.
4. Long Wenjun; Zhang Ying; Wang Jiaxing. Realistic interpretation and path selection of rural cultural revitalization [J]. Agricultural Economic Issues, 2019 (12): 15-20.
5. Fan He. "Art Intervention" Makes Rural Revitalization "Artistic Flower Road Release" [J]. Journal of Management Cadre College, Ministry of Agriculture and Rural Affairs, 2018 (4): 63-65.

Research on the Judicial Expertise of Electronic Evidence Based on Blockchain

Bo Zhang

Hainan Vocational College of Political Science and Law, Haikou, Hainan, 570100, China

Abstract. Give full play to the advantages of blockchain technology, guarantee the consensus of public credit institutions with the help of hierarchical nodes, and form a judicial alliance chain, so that the public can participate in the data verification activities of electronic evidence in the form of organizations, promoting the socialization and development of judicial activities. By using the credible time-stamp evidence, the judicial block chain presents the extremely formidable application superiority, the node verification based on the block chain consensus mechanism, is helpful to form the block chain judicial system which has the extremely application prospect. Through the analysis of the practical application and application prospect of the judicial block chain, the author makes a complete analysis of the idea of the electronic evidence judicial appraisal based on the block chain.
Keywords: Blockchain, electronic evidence, forensic science, Applied Research.

The "participatory" judicature under the block chain structure makes the public become the joint verification node, which makes it possible for the society to share the risk of judicial perjury, this kind of brand-new judicial mode shows the bright banner of the rule of law of socialism with Chinese characteristics in the new era through the institutionalization of public participation in the national governance mechanism.

1 Definition of blockchain evidence

In the new judicial environment, a large number of civil, criminal and administrative cases are also involved in electronic evidence. Facing the severe reality, in recent years, China has accelerated the revision of the relevant laws and confirmed the electronic evidence as a new form of evidence, thus laying a solid foundation for effectively combating crime. Strictly speaking, the blockchain itself is not evidence, but a way to store, carry and fix the electronic data evidence. At the legal level, article sixty-three of China's civil procedure affirms the legal effect of "electronic data" as evidence in civil proceedings. On this basis, in October 2019, the newly revised regulations on Supreme People's Court evidence in civil proceedings (hereinafter referred to as the regulations on evidence in civil proceedings) detailed the scope, judgment methods and admissibility standards of "electronic data" evidence, although there is no mention of "blockchain" evidence as a clear classification of "electronic data" evidence, but article 14"other information that can prove the facts of the case stored, processed and transmitted in digital form" provides support for the application of new information technologies such as "blockchain" in civil litigation evidence.

For the first time, the data on the blockchain is clearly defined as the type of evidence in civil proceedings, is in 2018, the entry into force of the Supreme People's Court on the Internet Court to hear the case of a number of issues (hereinafter referred to as the "Internet Court Rules"). Eleven of them is clear, "the Internet Court shall confirm the authenticity of the electronic data submitted by the parties, which can be proved by such technical means as electronic signature, trusted time stamp, hash value verification, blockchain and other evidence collection, fixation and tamper-proof, or by the authentication of the electronic evidence collection and deposition platform. "blockchain evidence" refers to the electronic data evidence that can be effectively tamper-proof, authenticity, and provable through the use of blockchain technology or through the use of the technology's evidence platform to collect, store and fix facts. At present, this kind of evidence has been accepted by judicial interpretation in our country, and it is widely used in the case of Internet court.

The reason why blockchain data can be used as litigation evidence is that the blockchain itself is a way to carry electronic data, on the other hand, it is based on the basic principle of blockchain. According to the white paper on the use of blockchain forensic evidence, it is generally considered that blockchain is a kind of bookkeeping technology, also called distributed bookkeeping technology, which is maintained by many parties, uses cryptography to ensure the security of transmission and access, can realize the consistent storage of data, is difficult to tamper with, and can prevent repudiation. The so-called "blockchain technology", can be understood as a series of existing technology combination application, mainly including: P2P network. Different from the traditional centralized network structure, P2P network refers to the peer-to-peer status of each node in the system, can act as a server. The resources and services in P2P network are scattered on all nodes, and the information transmission and service implementation are directly between the nodes. Consensus mechanism. Block chain authentication relies on consensus algorithm to achieve consensus rules, node selection and data consistency verification. The independent nodes of the phase chain can send, read and transmit data information to each other. P2P network and consensus mechanism also constitute the decentralization and information disclosure of blockchain technology. Digital signature technology. Digital signature technology uses public-key cryptography to encrypt (form signature) , decrypt operations, and verify that the message forming the signature has not been tampered with. Therefore, digital signature technology can verify the integrity of information. Chain storage. Chaining is a linked list constructed by cryptographic algorithms such as Hasche pointers, and the latter block contains the characteristic information of the previous block, so if you want to modify the data in one of the blocks, then the sequential block data in the chain-like storage structure must be completely modified. The storage structure effectively improves the difficulty of modification and the ability of anti-tamper and anti-forgery of the stored data. In addition, the blockchain authentication platform also uses other information technologies, including trusted timestamps, secure storage, cloud services, and so on. These technologies, together with blockchain technology, ensure the efficient and safe operation of the platform system.

With the help of blockchain technology, litigants can record the whole process of submitting electronic contracts and protecting their rights in the chain of electronic evidence, each organization node carries on the whole process tracking. The platform can verify the contents of certificates, electronic signatures, party information and hash values. Open the "judicial electronic evidence cloud "electronic evidence preservation and authentication certificate"(hereinafter referred to as the certificate) to verify the identity of the parties, confirmed that the parties have passed the real name authentication. Through the multi-party evidence verification platform, the electronic data submitted by the parties is confirmed to be the electronic data in the certificate, thus ensuring the originality of the electronic data. The recognition of blockchain evidence can not be separated from its own technical characteristics.

2 The effective path of electronic evidence judicial expertise based on blockchain

The rapid development of the information age has not only brought culture shock to our life, but also changed our life rhythm, and with the increasing number of cases of information technology crime and the diversification of criminal means, electronic evidence emerges as the times require, and the collection of electronic evidence has become the focus of providing important clues and solving cases. As for the judicial determination of blockchain evidence, it should closely conform to the civil procedure and the relevant judicial interpretation, and neither should the standard of evidence determination be raised because it belongs to a new and complicated technical means, also can not give up because this technology has the characteristic that tampers with hard, delete to set necessary standard. The Internet court rules, which came into force in 2018, focus on the authenticity of blockchain evidence, and the Supreme People's Court 2021 on January 21 released the draft regulations on certain issues concerning the online handling of People's Court cases (draft for soliciting opinions)(hereinafter referred to as draft for soliciting opinions) , between fourteen and 17[7] , the standards for the judicial determination of blockchain evidence are proposed in terms of the validity of blockchain evidence, the rules for the examination of blockchain evidence, the examination of the authenticity of data prior to the uplink of blockchain, and the determination of the reinforcement of blockchain evidence.

Blockchain technology is constantly integrated into the field of enabling justice. China's judicial practice departments have already explored in various fields, and have also formed a system platform with judicial practicality, for example, "block chain evidence" chain, "block chain appraisal" chain, "Block chain notarization" chain, "block chain arbitration" chain, "judicial chain intelligent contract" to enhance the binding force and enforcement of pre-litigation mediation agreement, and the local department of justice to explore the "block chain legal services" chain around the rule of law and the distribution of judicial administration. To verify the creditability time in the certificate verification of the multi-party evidence verification platform, after submitting the trusted time credential number and hash value (digital fingerprint) verification, the result of "time and event correspondence are clear, and time is certified as authentic, authentic, authentic, and authentic from the Time in Taiwan" is obtained, to confirm the parties to submit electronic evidence of a reliable source of time, traceable and verifiable. The platform supports functions such as evidence verification, electronic evidence storage, electronic evidence collection, blockchain public notice, etc. , focus on the settlement of Internet copyright, internet finance, electronic contracts and other areas of dispute types. In order to reduce the cost of judicial proceedings, improve the efficiency of dispute resolution, and prevent data tampering, the data of the internet evidence platform and the block chain established by the Court should be merged and shared in the stage of trial and execution of cases, we will promote litigation services and the establishment of a public trust system.

3 Suggestions on the improvement of electronic evidence extraction in judicial practice

The electronic evidence appraisal is a new type of judicial appraisal in the judicial practice, which is facing two important issues: technical improvement and legal regulation. In the aspect of legal regulation, it mainly involves the core issues such as the definition of electronic evidence identification, the scope and types of electronic evidence identification, the basic procedures, technical norms and basic principles of electronic evidence identification. In the Digital Age, the judicial trust system needs two aspects to strengthen: strengthening the tamper-proof of the judicial information, enhancing the transparency and supervisability of the judicial procedure. The integration of blockchain smart contracts can change the way judiciary collaborate and interact with information, and enable data sharing in an automated and intelligent, holistic and collaborative manner, this can not only reduce the uncertainty, randomness and complexity caused by human factors, but also improve the efficiency of judicial cooperation and rapid response capacity. The essence of blockchain is a new type of database software integrated with distributed network, encryption technology, intelligent contract, etc. , it promises to solve trust and security problems in cyberspace. From the technical point of view, blockchain is not only a technical innovation, its wide application also means a change in governance, is expected to develop into a new information infrastructure to realize value interconnection, and rebuild social trust pattern, this will revolutionize the industry and create new forms of business.

Block chain evidence and its rule preliminary formation. The scope of application is not limited to civil and administrative cases, but also includes criminal cases of speedy trial procedure, commutation and parole cases, as well as criminal cases that are unsuitable for offline trial for other special reasons. This not only means the integration of block chain technology and litigation rules, but also opens the way for the development of block chain judicial model.

As an information technology, blockchain is the result of human intelligence and human behavior. Therefore, the law on blockchain refers to a set of behavior rules system which is created and applied by human beings. So cross-blockchain is like using the Internet, with huge initial dividends, especially in areas of multi-disciplinary integration. Whether it is Blockchain + law, blockchain + AI, or Blockchain + any industry, how to combine the blockchain with the original industry has become a matter of great concern to everyone. This year, whether looking globally or at our country, the area of blockchain and digital currency has entered a period of rapid development. While the technological basis and business model are constantly innovating, the trend of legal policy and supervision is becoming clearer, it also provides an important foundation for the further application of blockchain technology.

The platform supports functions such as evidence verification, electronic evidence storage, electronic evidence collection, blockchain public notice, etc. , focus on the settlement of Internet copyright, internet finance, electronic contracts and other areas of dispute types. In order to reduce the cost of judicial proceedings, improve the efficiency of dispute resolution, and prevent data tampering, the data of the internet evidence platform and the block chain established by the Court should be merged and shared in the stage of trial and execution of cases, we will promote litigation services and the establishment of a public trust system.

At present, in the judicial cognizance of blockchain evidence, the most concerned problem is the judgment of the authenticity of evidence. According to the current judicial precedent, the court generally reviews the facts concerning blockchain evidence from two aspects, namely, the conformity of the platform of blockchain evidence and the authenticity of blockchain evidence. When using the blockchain authentication platform for fact fixing, should choose a well-known platform with important qualifications, such as block chain information service registration and filing, etc. , and CA certificates issued by state-authorized third-party electronic certification bodies, etc. , the authenticity of pre-deposition evidence should be ensured as much as possible.

4 Conclusion

Under the condition that the rule of block chain evidence is not clear enough, the authenticity of block chain evidence can also be proved by applying for judicial expertise. Of course, the ultimate goal of blockchain evidence is to provide a convenient, tamper proof bearing and fixing way for the People's Court to judge and restore the facts by means of decentralization. As for the rule and standard of the block chain evidence, the detailed standard should be issued as soon as possible, so as to be better applied in the judicial practice.

5 Acknowledgements

High-level Talent Project of 2021 Hainan natural science foundation, "research on electronic evidence forensic examination based on blockchain" project number 621RC612(Junko [2021]253)

References

1. Zhang Yujie. The judicial application, system problems and evidence law innovation of blockchain technology [J]. East Law, 2019, (3): 99-109.
2. Zheng Guan, Fan Ketao, Wu Hong. The path to determine the authenticity of blockchain electronic evidence [J]. People's Justice, 2020, (4): 106-111.
3. Bai Xuetong, Wang Qiuyun, Chen Ying. Research on the Application of Electronic Evidence Platform Based on Blockchain Technology [J]. Cyberspace Security, 2020, (10): 104-109.
4. Chen Wei, Zhang Mingyang. The application and development of blockchain in Internet justice - an empirical analysis based on the judicial blockchain platform of Hangzhou Internet Court [J]. People's Justice, 2020, (31): 4-8.
5. Yi Ran. The application exploration and practice of blockchain technology in the judicial field - an empirical analysis based on the Beijing Internet Court Tianping Chain [J]. China Applied Law, 2021, (3): 20-34.

Research on the Influence of Marketing Information Quality on Consumer Behavior -- Taking Wechat Marketing as an Example

Zhuoye Zhang[1], *Sixue* He[2]

[1]Software Engineering Institute of Guangzhou, Guangzhou, China, 510990
[2]Guangzhou College of Technology and Business, Guangzhou, China, 510850

Abstract. With the continuous prosperity and development of Internet technology, the amount of information generated is increasing day by day, which inevitably leads to a series of problems such as marketing information surplus and user perception disorder. How to improve the quality of information and reduce the information disturbance to users has become an urgent problem for social media. This paper takes wechat marketing as an example, through literature research, questionnaire collection, SPSS statistical analysis and Amos structural equation fitting, based on previous research and the development status at home and abroad, summarizes various dimensions of marketing information quality, and combines drive theory and relationship marketing theory to identify three levels that affect user satisfaction and behavior: information content, utility and carrier quality. After reliability and validity test and structural equation fitting analysis, it is found that in three levels, there are five dimensions (authenticity, integrity, incentive, security, system performance) with user satisfaction as the intermediary variable, which positively affect user consumer behavior, and the influence coefficients of each dimension on user satisfaction are obtained. Among them, the influence coefficient of incentive on user satisfaction is the largest Suggestions on strengthening users' interest perception, improving the richness of media, optimizing the system layout and improving the information security mechanism.

Keywords: Wechat Marketing, Structural Equation, Consumer behavior, User Satisfaction, Information Quality.

With the development of the economy and information-based society, the rapidly growing amount of social media marketing information has made consumers feel confused and bored, which leads to search barriers and the decline of information marketing effectiveness [1]. How to meet users' demand for marketing information with high quality and actively stimulate users' sharing value as network nodes has become an urgent task to be solved in social marketing. On the basis of previous studies, this paper summarizes the influencing factors of marketing information quality and tries to analyze how these factors affect user satisfaction and user behavior.

1 Literature review

1.1 The definition of wechat marketing and wechat marketing information

Wechat marketing refers to the expression and shaping of various product forms and brands by enterprises relying on wechat platform, and the one-way continuous output of users [2]. While wechat marketing information is the manifestation of enterprises, brands, products and other information on wechat, aiming to build a long-term good relationship with users.

1.2 Domestic and foreign research review

(1) Research on marketing information based on cognition

The IS system model provided by Delone and Mclean IS the basis of this study. When studying IS system, they believe that system quality and information quality jointly affect user satisfaction and usage, and ultimately affect individuals and organizations. The specific model IS shown in Figure 1 [3].

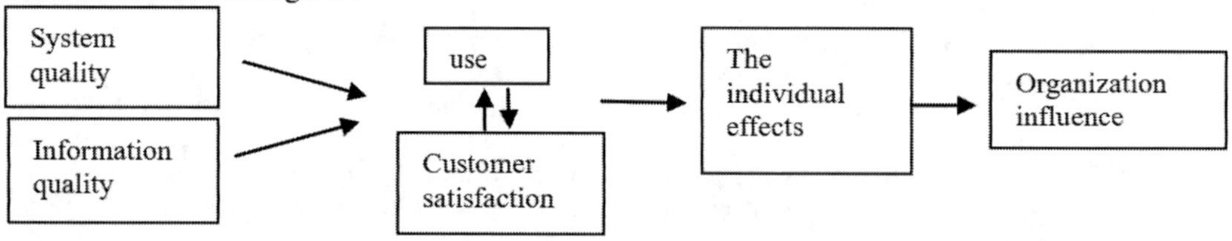

Fig.1 Success model of Delone information system.

Urbach N, Smolnik S & Riempp G (2009) conducted in-depth research and analysis on the status quo of the success of information systems from 2003 to 2007, and confirmed that IS success model IS still the main basis of IS success measurement [4].

Domestic researches on the quality of marketing information are mostly concentrated in Weibo, online communities, wechat public accounts, etc.

Feng Xuyan (2015), taking consumers as the subject, studied the influence of consumers on the acceptance intention of wechat marketing, and concluded that the ease of use, usefulness and pleasure of marketing can positively affect users' attitudes, while the perceptual interference has a significant negative impact on users' attitudes. Hu Yuan (2017) through the microblog information quality and information source credibility of fuzzy hierarchy evaluation model analysis, the influencing factors on the quality of the information from large to small is arranged in the order: information timeliness, information effect, authority, information value, information accuracy, information correlation, information content, information presentation form [5].

In conclusion, perceived comfort can positively affect users' attitudes, while interference can negatively affect users' attitudes.

(2) Research on marketing information based on user behavior

Information sharing is an expression of user behavior, which plays an end role in the process from a series of emotional changes caused by information contact to the final acceptance and sharing of information. For enterprise marketing, it is an ideal situation for users to share information to realize brand communication and chain spread of brand information on social media.

Song Weixiang and Jia Jia (2019) confirmed that information content quality, information form quality, information utility quality and information carrier quality have a positive impact on users' interactive behavior through small-scale interview method and user behavior data analysis of public accounts .

Chen Xing, Zhang Xing, Xiao Quan (2019) studied users' willingness to continue knowledge sharing in online health communities, took user satisfaction as an intermediary variable, explored the influence of other factors on the willingness to continue knowledge sharing through satisfaction, and found that there was a significant positive impact.

By summarizing relevant studies at home and abroad, it can be found that user behavior has a certain purpose, that is, there is a formal motivation in user behavior. For information itself, internal and external incentives have an impact on user behavior.

2 Research model and hypothesis

2.1 Research model

In this study, the maturity scale of predecessors was used as a prior scale to construct a model of information quality affecting user behavior. User satisfaction was added as an intermediate variable, and the second-order structure of the scale was set. Information quality was divided into three first-order factors, content quality, utility quality and carrier quality. There are seven dimensions of authenticity, expressiveness, timeliness, motivation, interactivity, security and system performance as second-order factors. See Figure 2.

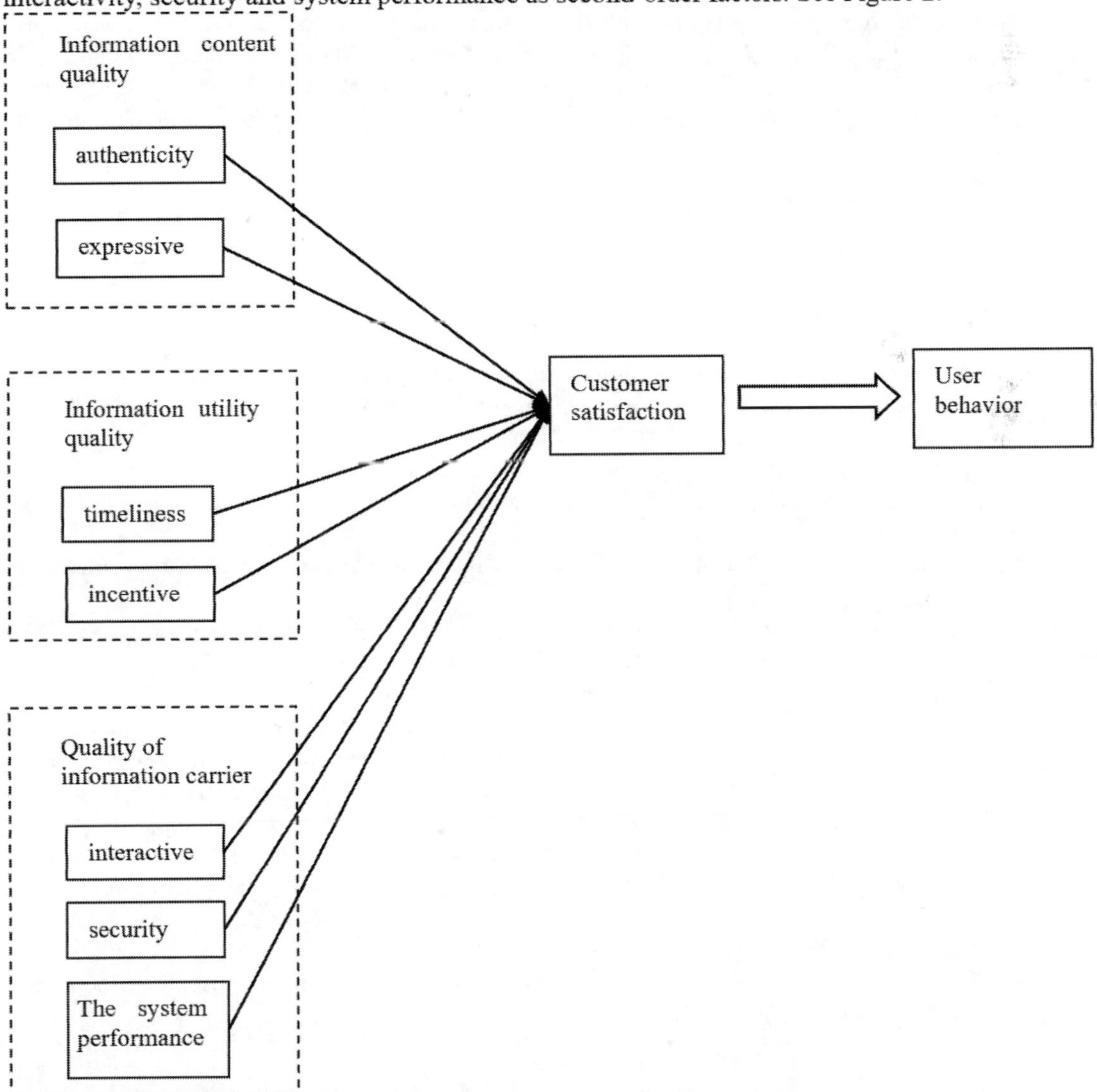

Fig. 2 Influence model of information quality on user behaviour.

2.2 Research hypothesis

(1) Information content quality related assumptions

For consumers, the authenticity and reliability of information is the most basic. Once information becomes false and unreliable, consumers' cognition of information will stop there. However, incomplete information will lead to distortion of part of the content. In the long run, it will lead to deviation between users' expectation and actual perception, thus changing users' behavior and attitude.

Based on this, the following assumptions are made:

H1 The higher the authenticity of information, the more satisfied consumers are with the quality of information;

The higher the integrity of H2 information is, the more satisfied consumers are with the quality of information.

(2) Relevant assumptions of information utility quality

Timely and applicable information is one of the indicators for users to obtain effective information. Information has significant timeliness. The current and constantly updated information, timely and effective information can help users make decisions. The more effective information is for users to measure decisions, the more valuable it is. When cao Ruichang (2002) studied information quality and its evaluation index system, he pointed out that information has an obvious time limit, beyond which information will lose its value.

Based on this, the following assumptions can be made:

The higher the timeliness of H3 information is, the more satisfied consumers are with the information quality.

The higher the incentive of H4 information is, the more satisfied consumers are with the quality of information.

(3) Relevant assumptions about the quality of information carrier

Studies have shown that information content that can relate to users' interest demands can better stimulate user participation and promote the formation of information influence within the group .

Based on this, the following assumptions can be made:

The higher the interactivity of H5 information carrier is, the more satisfied consumers are with the information quality.

The higher the security of H6 information carrier is, the more satisfied consumers are with the information quality.

The higher the system performance of H7 information carrier is, the more satisfied consumers are with the information quality.

(4) Assumptions about user behavior

When studying the influencing factors of community e-commerce user information and user behavior, it is found that a good interaction between users and the platform can greatly increase users' favorable impression on the platform, and the simplification of platform functions, the aesthetics of the interface and the overall layout of the interface can give users an intuitive visual impact . Therefore, this paper argues that consumers' decision to adopt user behavior depends to a great extent on their personal attitude at the moment of reading information.

Based on this, the following assumptions can be made:

H8 users' satisfaction with enterprise marketing information positively affects user behavior of marketing information.

3 Questionnaire design and empirical analysis

3.1 Research variables of questionnaire design

Questionnaire design will follow the following 3 steps.

(1) Formation of initial scale: The existing maturity scale was quoted to improve the overall fitting degree and recognition degree of the scale.

(2) Questionnaire pretest: The initial scale was put into some target groups for questionnaire pretest to test the reliability and validity of the overall data of the questionnaire and the model, and to test the fitting degree of the questionnaire model according to the factor test.

(3) Forming the final questionnaire: Appropriate deletion and modification should be made according to the results of the predictive tests to ensure that the scale has good aggregation effect and discriminating validity.

3.2 Questionnaire predictive test

In the predictive test, the author adopted the convenience sampling survey method, and selected undergraduates of The Hua Soft Software College of Guangzhou University as the sample samples. These students have certain thinking ability and have experience in user behavior of marketing information of relevant enterprises (such as Pindoduo, Tmall, Taobao Double eleven, jingdong, etc.). 84 questionnaires were collected, and 70 were valid after screening conditions.

Reliability test is the reliability test of the questionnaire data, reflecting the stability of the data, and is an evaluation index of the truthfulness of the questionnaire data. According to relevant standards, the coefficient 0.7-0.8 indicates that the data has considerable credibility, 0.8-0.9 is very good, and above 0.9 is very good.

3.3 Empirical analysis

In the formal investigation stage, the questionnaire was distributed to all users of wechat in the whole network for two weeks, and was spread by using crony sampling. Finally, 578 questionnaires were collected, of which 538 were valid, with an effective recovery rate of 93.07%.

(1) Descriptive statistical analysis

After summarizing and analyzing the collected questionnaire data; 39% are male and 61% are female; The age distribution is mainly between 19 and 35 years old, mainly young and middle-aged; The occupation distribution is more uniform, the education level is generally college or above.

(2) Reliability analysis

The effective sample data were imported for reliability test. The relevant results were shown in Table 3-2, $\alpha=0.873>0.8$. The reliability of each measurement dimension was measured by stratification. The Alpha value of each subitem is lower than the overall α value, indicating that the questionnaire has good reliability in the overall structure.

(3) Validity analysis

Validity analysis was carried out in SPSS 21.0, and KMO=0.855>0.8, Bartlett's value 8023.4, sig=0.000<0.001 were obtained, indicating that the data could be further exploratory and confirmatory factor analysis.

(4) Factor analysis

After validity analysis, principal component analysis and rotation were adopted, and it was found that there were 9 variables that could be extracted, with explanatory degree =74.664%>70%. Therefore, this study believed that the explanatory degree of these 9 factors was acceptable for all data.

Rotation of the component matrix and cancellation of the absolute coefficient less than 0.6 shows that each factor subitem has a high load on the same factor, and factors 1-9 correspond to user behavior, security, incentive, system performance, integrity, timeliness, interactivity, authenticity and user satisfaction respectively.

CFA analysis was conducted for a total of 9 variables and 30 sub-items. AVE values of all variables are greater than 0.5, and CR values are higher than 0.7, indicating that the data has good convergence (convergence) validity.

4 Management suggestions

4.1 Timely strengthen user behavior

This study shows that incentive information has the greatest influence on user satisfaction and user behavior. Reinforcement theory holds that timely positive reinforcement of desired behavior will improve the probability of occurrence of the behavior. At present, for enterprise marketing information, material and spiritual stimulation after the information can attract users to click, participate in and forward to a greater extent.

4.2 Enhance the richness of media

The integrity of information expression can make users' perception and actual performance tend to be consistent, while rich expression can enhance the interest of information and increase its attractiveness. WeChat compared to other social platforms, an obvious advantage is a rich variety of media, the enterprise can through the media to define products, at the same time, interesting, for enterprise marketing information, in addition to common words, pictures, can also through some specific conditions to realize the information acquisition, such as sliding interface, AR scanning, Mobile phone rotation and other ways to enhance the richness of the media, increase the interest of information, to attract users' attention.

4.3 Optimize the system layout

The functionality that the user needs should be set in an obvious location, and never add an intermediary for the user to achieve the goal.

Therefore, the design of functional modules should be established on the basis of good interaction between enterprises and users, otherwise not only increases costs, but also damages users' trust in the brand image.

Conclusion

Through the actual questionnaire survey and data statistical research methods, the following conclusions are drawn: wechat marketing information affects user satisfaction and produces user behavior in five factors, and the order of the influence coefficient is motivation, system performance, integrity, authenticity and security.

References

1. Fang Tiankun, Yang Lixue, Zhao Yingying. The influence of information quality and risk taking on consumers' purchasing intention in wechat marketing [J]. Foreign Trade

and Economic Cooperation, 2015(11):42-44.

2. Li Y. Research on wechat viral marketing factors and their impact on consumers' willingness to share [D]. Guangzhou: Jinan University, 2016.
3. De lone W H, Mclean E R. Information systems Success: Journal of Management Information Systems, 1992, 29(4):7-62.
4. Urbach N, Smolnik S & Riempp G. Thc State of Research on Information Systems Success[J]. Business & Information Systems Engineering, 2009, 1(4): 315-325.
5. Feng X Y. Research on factors influencing consumers' acceptance intention of wechat marketing [D]. Beijing University of Posts and Telecommunications, 2015.

Logistics Transportation Time Optimization Based on Fuzzy Particle Swarm Optimization

Bingqing Zhang

Guangdong Provincial Transportation Planning and Research Center, Guangdong, Guangzhou 510101, China

Abstract. The fuzzy particle swarm optimization (FPSO-RRB- algorithm is used to solve the logistics transportation time optimization problem, a three-dimensional particle coding method based on receiving point, particle position sequence and particle position rounding operation is proposed, the results are compared with those of fuzzy particle swarm optimization algorithm. The experimental results show that the fuzzy particle swarm optimization algorithm can effectively optimize the logistics transportation time.
Keywords: Fuzzy particle swarm optimization; logistics and transportation; time optimization; coping strategies.

The logistics of the future should be data-driven and intelligently optimized. By means of a hybrid algorithm, combining the advantages of each algorithm and making up for the shortcomings of each algorithm, a more efficient and feasible algorithm scheme is gradually formed, which is widely used in the optimization of supply chain network, the sorting of orders in a warehouse, and the balance between ordering and demand, optimization of distribution path and other fields. Because many logistics problems are similar, through in-depth study to use the appropriate algorithm to solve these problems.

1 Logistics transportation time optimization background based on fuzzy particle swarm optimization algorithm

Logistics Center is a place or organization engaged in logistics activities, can handle all kinds of transportation, storage, packaging, loading and unloading, circulation processing and other logistics operations, is the logistics information processing point. In today's supply chain management, logistics center as a link up and down the supply chain, played a vital role. Reasonable logistics center location can effectively save costs, ensure the efficient operation of the logistics system, supply chain management of strategic significance. The logistics transportation time problem is always the research domain which attracts people's attention, already many scholars have carried on the research to the location problem, proposed many kinds of methods, such as Baumolwolfe, Barycenter, Tabu Search, fuzzy particle swarm optimization, simulated annealing, analytic hierarchy process and artificial neural network. Because saving transportation cost and shortening transportation time are the primary problems that should be considered in the location of logistics center.

In the long-distance logistics transportation, the distance between two places can be considered approximately as a straight line distance, and this estimation method becomes more effective as the distance between two places increases. The common methods of environment modeling are visualization method, grid method and topology method. Because the topology method is simple in use, the environment modeling is easy, and it is more suitable for the transfer between cities, therefore, the topological method is used to model and simulate the environment. The mutation rate of the population is calculated by using the adaptive mutation probability and the secondary mutation probability, and the corresponding habitat is selected for mutation operation according to the adaptive mutation probability and the secondary mutation probability to update the optimal solution; If the number of iteration is reached, the optimal solution, that is, the optimal transportation scheme, will be output; otherwise, the iteration will proceed.

This paper presents a vehicle routing optimization method based on block matrix and fuzzy transportation time, which considers the selection of two transportation modes under the condition of fuzzy transportation time, thus, the optimal transportation scheme under uncertainty can be obtained. In order to achieve the above objectives, a logistics transportation time optimization method based on fuzzy particle swarm optimization algorithm is proposed, it includes the following steps: establishing the route optimization model based on fuzzy transportation time, circulation transportation sequence and cross distribution center assignment; initializing the solution of the route optimization model, I. E. Habitat, the block matrix is used to represent each habitat, the fitness index value of each habitat is calculated, and the initial optimal solution is determined according to the fitness index value; And judging whether the habitat is to be moved in or out according to the immigration rate and the emigration rate, and if the immigration or emigration operation is needed, the vector substitution is performed on the block matrix corresponding to the habitat to realize the immigration or emigration operation.

Vehicle routing problem (VRP) refers to a logistics distribution center that supplies goods to a certain number of demand points with different demand for goods, and on the basis of satisfying the demand points' distribution requirements, carries out reasonable route planning, finally, it achieves the goal of the shortest transportation distance and the lowest transportation cost. Because the study of this problem is of great practical significance, the heuristic has great advantages in solving the vehicle routing problem, the study of heuristic has also been enriched. Time windows and fuzzy constraints are added to the short distance open vehicle routing problem (VRP) , and the hybrid ant colony algorithm is used to solve the VRP, the problem is solved by a heuristic with a variable domain search strategy, which solves the vehicle routing problem with backhaul and time windows, and by a hybrid particle swarm optimization algorithm.

2 Logistics transportation time optimization scheme based on fuzzy particle swarm optimization algorithm

Particle swarm optimization, proposed by Kennedy and Eberhart in 1995, is an evolutionary computing technique derived from simulations of social behavior, such as bird and fish predation. The PSO algorithm randomly produces an initial population and gives each particle a random velocity. During the flight, the velocity and trajectory of the particles are dynamically adjusted according to their own flight experience and that of their companions, the whole group has the ability to fly to a better search area. Particle swarm optimization has been widely used in function optimization, neural network training, fuzzy system control, pattern recognition and engineering applications, it is proved that the optimal solution can be obtained with less computational cost.

Particle swarm optimization (PSO-RRB- is applied to solve the location problem of logistics center. In a 2-d search space, the population is made up of M particles, where the position vector of the I particle represents the feasible address coordinates of a logistics center on the plane. The fitness value fi of the first particle is calculated from the logistics cost formula of the objective function of the location problem. The first particle evolves to the individual optimal position found in the G generation, and the whole particle swarm iterates to the global optimal position found in the H generation. From generation H to generation H + 1, the velocity and position of the I particle in the j-dimensional subspace can be adjusted.

As an optimization, the route optimization model based on fuzzy transportation time, circulation transportation sequence and cross distribution center assignment is established, and the optimization goal is to maximize the satisfaction of the manufacturer, s is the fitness function of vendor satisfaction, G is the vendor set, G is the vendor G, K is the vehicle set, K is the vehicle k, rKG is the manufacturer G's satisfaction and the manufacturer's order is transported by the circulating transport vehicle K, WKG is the weight of the manufacturer G's satisfaction and the manufacturer's order is transported by the circulating transport vehicle k, wLG is the weight of manufacturer G's satisfaction and the manufacturer's order is transported through the cross-distribution center L, S is the supplier set, I is the supplier I, [CG, DG -RSB- is the time window required by manufactuG.r g, dig is vendor G's order to vendor I, and DG is vendor G's order to all vendors.

The YIKG and xilg in the objective function are the decision variables, and the BKG1, BKG2, and BKG3 in the objective function are the fuzzy time of arrival of the order from the manufacturer G that is transported by the circulating transport vehicle k, the fuzzy time is the fuzzy number of BKG, and BKG is the time when the vehicle K reaches the manufacturer G through circular transportation. In the formula bkg = BKI + stki + tig, k ∈ K, g ∈ G, I ∈ S, bKI is the time when the circular transport vehicle k arrives at supplier I, stki is the service time of the circular transport vehicle K at supplier I, TIG is the transport time from supplier I to manufacturer G, stki = γdigyikg, γ is the loading or unloading time per unit product, Dig is the order from G to supplier I, and YIKG is the decision variable FLG1, FLG2, and FLG3 in the objective function are the fuzzy times of order arrival for the manufacturer G through the cross-distribution Center L, which is the fuzzy number of FLG, fLG is the time for the vehicle to arrive at manufacturer G through the cross-distribution Center L, FLG = ALG + TLG, L ∈ L, G ∈ G, and ALG is the time for the Order of manufacturer G to leave the cross-distribution center L, tLG is the transportation time from the cross-distribution Center L to the manufacturer G, γ is the loading or unloading time per unit product, tilg is the initialization time of the vehicle transportation to the manufacturer G through the cross-distribution Center L, q is firm Q, DIQ is firm Q's order to supplier I, xilq and xilg are decision variables, and til is supplier I's shipping time to cross-distribution center L, dig's an order from supplier I for firm G.

The particle swarm optimization (PSO-RRB- coding is similar to the fuzzy particle swarm optimizatiFPSO-RRB-pso) double integer Theing. the number above the first layer segment indicates which vehicle the customer will be assigned to, the greatest advantage of the representation method is that every customer demand point can get the vehicle delivery service, and the demand of each demand point can only be completed by a certain vehicle, so that the calculation of the feasible process of the solution is greatly reduced.

3 Logistics transportation time optimization experiment based on fuzzy particle swarm optimization algorithm

In the selection operation of fuzzy particle swarm optimization, the first step is to calculate the fitness value of each chromosome, the second step is to calculate the proportion of the total, and the third step is to sort according to the fitness value from large to small.

According to many research theories, the computer uses the algorithm to process the data, the sort time accounts for 25% of the total processing time, then enhances the sort time to be able to enhance the algorithm processing efficiency. In this paper, it is reasonable to introduce fast sorting into the selection operation of fuzzy particle swarm optimization algorithm, and there is no conflict. It can not only improve the ranking time of fuzzy particle swarm optimization algorithm, but also improve its overall time.

The analysis of the efficiency of quick sort shows that it is still the best method with the lowest time complexity by many theories, experiments and comparisons. When the record is large, then should choose fast sort, from the average time performance point of view, other sort slightly inferior to fast sort and merge sort, these two methods when the input size is large, fast sort will prevail. Fast sorting is characterized by extremely fast sorting, less data movement, and eventually make the sequence from small to large sorting. Fast sorting algorithm is currently the most practical algorithm, is very suitable for dealing with complex problems, when the scale of the problem is larger, the execution time of using the algorithm is longer, compared with other algorithms, the efficiency of using fast sorting is higher, it's even more effective. The efficiency of the algorithm is usually measured by time complexity and space complexity, and the running time of the algorithm is obtained by comparing the time complexity of the algorithm under the premise of guaranteeing the same storage space, that is, S (N) .

In view of the characteristics of the optimization problem of logistics distribution route, a genetic algorithm is constructed to solve this kind of phenomenon, which includes the selection of the most classical fitness function with simple design, and the introduction of the fast sorting method in the selection operation, the main reason is that the complexity of genetic algorithm has been reduced. Select the natural number of the code, 0 as the starting point (distribution center) , every other integer for each customer point. Suppose there are 10 requirement points, 1 to 10 representing 10 different requirement points, and by starting with 0 and ending with 0, when a vehicle is delivered, it leaves the distribution center and returns to the distribution center. K distribution paths are randomly generated from each demand point, and then k distribution paths constitute the initial population, the size of the initial population is K.

The principle of proportional fitness calculation is that the feasible solution of the problem is substituted into the model equation to get the target value. If the value is smaller, the fitness is better. This method is simpler than the sort-based fitness calculation, convenient. According to the data, the fitness function can be designed according to the need of solving the problem. There are many fitness functions at present. The selection and design of fitness function is as simple as possible to minimize the time complexity of calculation, the fitness function will affect the convergence speed of genetic algorithm and whether a satisfactory solution can be found, and whether it reflects the genetic algorithm "survival of the fittest" characteristics. According to the calculation method of fitness function, the first step calculates the fitness value of each chromosome, the second step sums all fitness, and then calculates the fitness of each chromosome as a proportion of the total, at last, the quick sort is introduced to replace the common sort, because the quick sort can reduce the time complexity of the genetic algorithm.

4 Conclusion

Fuzzy particle swarm optimization (FPSO-RRB- has strong computing power in path planning, and it can adapt to many complex terrain with strong global searching power, can be used to search the parameters of the fuzzy particle swarm optimization algorithm. The results of simulation and contrast experiments show that the algorithm can optimize the distribution route, reduce the distribution cost and improve the distribution efficiency.

However, this algorithm depends too much on the length of the road, and many problems need to be considered in practical engineering. In the future, it will be studied more carefully by introducing road complexity factor.

References

1. Zu Zhiwen, Li Qin. Mahalanobis Distance Fuzzy Clustering Algorithm Based on Particle Swarm Optimization [J]. Journal of Chongqing University of Posts and Telecommunications (Natural Science Edition), 2019(2): 279-284.
2. Chen Tianpei, Wang Yuhui, Wu Qingxian, et al. Three-dimensional path planning based on fuzzy logic particle swarm algorithm [J]. Electro-Optics and Control, 2020(6): 1-5.
3. Fan Zhimin, Wang Baolin, Fan Yu. Optimization research on fuzzy PID control of vehicle semi-active suspension based on particle swarm algorithm [J]. Machinery and Electronics, 2020 (11): 76-80.
4. Lin Xinyou, Wang Zhaorui. Adaptive multi-objective control strategy using particle swarm optimization algorithm to optimize fuzzy rules [J]. Control Theory and Application, 2021 (6): 842-850.
5. Wang Hangyu. MPPT algorithm combining particle swarm and fuzzy algorithm [J]. Chinese Journal of Detection and Control, 2021 (4): 112-116.

The Transformation and Upgrading of English Teaching Modernization under the Background of Information Technology

Lijuan Zhang

Shanghai Jianqiao University, Shanghai, 201306, China

Abstract. With the development of Information Technology, English teaching system has realized intelligent innovation and all-round development, especially by paying attention to students' learning demands and integrating all kinds of teaching resources, it realizes the essential optimization of English teaching ecology. Therefore, based on the mature application of information technology, it is of great practical significance to focus on the transformation and upgrading of English teaching modernization. Based on the trend of modernization and transformation of English teaching under the background of information technology, this paper puts forward some corresponding upgrading strategies and suggestions.
Keywords. Information technology; English teaching; modernization transformation; upgrading strategy

Most of the existing research focuses too much on technology and neglects the theoretical level of research. Compared with traditional foreign language teaching, how does information-based foreign language teaching draw on the research results of psychology, pedagogy, Neurology, Information Technology, second language acquisition and whether it is necessary to construct new theories and frameworks, how to promote the integration of educational technology and foreign language teaching is the key point to be discussed.

1 The transformation trend of English teaching modernization under the background of information technology

The use of information technology and the integration of digital resources in foreign language teaching should take into account various factors and links such as teaching objectives, teaching objects, teaching methods, teaching environment and teaching evaluation, etc. , at the same time, the teaching practice is guided by the theories of information technology, pedagogy, psychology and second language acquisition.

Defining the goal of information-based foreign language teaching. The first is the long-term goal of foreign language teaching. The ultimate goal of language teachers is to help students acquire knowledge and skills, develop autonomous foreign language learning, and become culturally aware global citizens in both real life and online communities. Setting long-term teaching goals will provide direction for the integration of foreign language teaching and information technology. At the same time, information technology should also serve the specific teaching objectives of each course, which requires foreign language teachers to consider "cohesion", such as how to link up the classroom, online and offline. Language learning outside the classroom or online is often unpredictable and disorganized, and teachers must consider how such learning fits in with specific classroom goals, as opposed to structured instruction in traditional classrooms.

Pay attention to the differences of learners and improve the information-based language learning literacy of learners. Researchers, teachers, and technologists need to understand learners' language proficiency, learning experiences, and learning expectations as a basis for developing, selecting, and adjusting technologies and tools. At the same time, there is a need to develop learners' Digital Literacy, students' proficiency and acceptance of emerging technologies does not mean they can use them critically, and teachers need to provide guidance and training, to help them combine technology and language learning effectively. To clarify the advantages and disadvantages of various teaching methods, especially the characteristics, advantages and limitations of various information technology and digital tools. Therefore, English teachers themselves should receive education information technology training to improve their information literacy. Information technology is constantly changing, full of unknown and various possibilities. Foreign language teachers need to take an open, flexible and critical attitude towards information-based foreign language teaching and learn how to teach safely and ethically, critical use of information technology.

Decision-making in combination with the actual teaching environment. Teachers should be familiar with the technical, linguistic and cultural resources available in a real teaching environment, and should be familiar with traditional curriculum arrangements, teaching scales, examination pressures and other factors, adjust the degree and form of informatization accordingly. This paper systematically evaluates the effectiveness of information-based foreign language teaching, including determining the evaluation contents and standards, selecting appropriate evaluation methods, and making teaching adjustments based on evaluation. For example, for online learning, on the one hand, you can use technology to obtain learning data for learner analysis. On the other hand, we can evaluate the learning effect through the form of learner's diary and learner's file.

2 The emphasis of English teaching modernization under the background of information technology

English class under the background of information technology can improve learning interest. The form of English teaching should be vivid, lively, intuitive and diversified, which is determined by the students' psychological characteristics, a great deal of diverse information, such as images, sounds and colors, act on students' multiple senses, which makes English teaching atmosphere harmonious, English learning interest strong, listening, speaking, reading and writing potential to be effectively developed. Interest is the best teacher, in a pleasant and relaxed learning atmosphere students easily take an active part in teaching activities, its potential learning requirements are stimulated. English teaching under the background of information technology can fully arouse the enthusiasm of students to move their ears, mouth and hands, combine concrete image with abstract language, let students be in the real world, and realize happy learning.

English class under the background of information technology can reduce the difficulty of learning. After all, classroom time is limited, the use of multimedia teaching can effectively use the limited teaching time and teaching space. Multimedia is an important means for teachers to highlight key points and break through difficulties. It can make students more receptive and make complicated knowledge more understandable. The use of multimedia can attract students' attention and interest in English knowledge. Different language environment provides students with a real situation for English communication. Students not only learn knowledge and deepen their understanding, but also can simulate the application according to the real situation, which not only breaks the teaching focus, it also makes learning English easy and easy.

English classroom under the background of information technology can promote learning interaction. The effect of interactive teaching is obvious. It is easy to realize teacher-student interaction and student-student interaction with the background of modern information technology. English is a basic subject with strong communicative, instrumental and practical nature, teachers should create a good teacher-student relationship, create a relaxed, harmonious and positive atmosphere, create a different language environment with the help of information technology, and provide students with as many opportunities to use language as possible, to realize language learning in situations, to master language in communication, to gradually realize the transformation of knowledge to skills, to build up students' self-confidence in learning, so that students at different levels have different degrees of harvest.

3 The content of modernization transformation of English teaching under the background of information technology

3.1 Vocabulary teaching

The first is the preparation of listening materials, the use of removable hard disk, U disk, memory cards, etc., to store the entire semester or all the school year's content, Plug and play, do not need to frequently change the content. These listening exercises can be played at any time, greatly saving control time, improving classroom efficiency, and content standards, no longer worry about deviation listening and reading, in the classroom you can hear the pure pronunciation of foreign language vocabulary. In order to stimulate the enthusiasm of learning vocabulary, we must first cultivate students' interest in vocabulary. The use of multimedia classrooms to play listening materials, the most direct visual, text, sound in front of students, the brain will receive information will soar and remember the frequency will be accelerated, over time, the improvement in your listening skills will increase by leaps and bounds.

3.2 Teaching of listening

Using information technology to broadcast listening materials, there are words, images and sounds, vivid, students' interest in learning hair-trigger. First of all, to deal with the key and difficult words in the listening materials, through the multi-media technology, the sound, form and meaning of the words are presented at the same time, fully mobilize the various senses of the students, and establish the direct contact between the various aspects of the vocabulary, design activities such as connecting lines, filling in blanks, options and substitutions, and intensify training. Secondly, prepare the background knowledge, using multimedia technology to display, intuitive, image of the knowledge background is easy to accept. With these building blocks, listening training is easier and more effective. English listening teaching under the background of information technology is rich and colorful, which arouses students' interest in learning and promotes their autonomous learning and active learning. With the development of English listening resources and the broadening of listening training channels, the content of listening teaching will be more abundant and the listening class will be more optimized.

3.3 Teaching of reading

Under the background of information technology, reading teaching has the characteristics of fast rhythm, large capacity, multi-level and high efficiency. The teacher prepares the lesson earnestly, prepares the multimedia courseware, presents the teaching content in time, displays the language activity material, presupposes the teaching link, constructs the thick reading classroom atmosphere. By arousing the enthusiasm of all kinds of sense organs, students can be stimulated to carry out reading activities. Before reading, through the text, images, pictures, animation and sound, and other forms of reading materials, and add relevant background information, so that students familiar with the reading theme, to predict the reading content, environment and vocabulary, make clear the reading purpose, and prepare for the development of reading activities. In reading, students are encouraged to take an active part in removing language barriers step by step, to understand the content of the text, to understand the logical order of the text, and to grasp the main idea. After reading, by presenting the main points, key words and the theme of the text, the students are guided to carry out role-playing, retelling the text and other communicative activities to consolidate the teaching content. Finally, through the courseware to display the relevant text and image information, encourage students to discuss and communicate, create real language communication situation, and cultivate students' comprehensive ability to use English.

4 The transformation strategy of English teaching modernization under the background of information technology

The language teaching under the information environment is different from the traditional foreign language teaching in the classroom. How to integrate various teaching methods with information technology, what principles should be followed, and what teaching models can be developed, how to combine with the teaching environment in China is a direction that can be explored. For example, project-based learning, cooperative learning, how to carry out task-based teaching in the information-based environment, how to combine game-based learning with Chinese teaching situation, and so on.

Information technology makes it easier for researchers to collect and analyze data. In the process of combining technology with foreign language teaching, learners' learning process, curve, pattern and behavior are worth paying attention to, and this kind of data can provide valuable enlightenment for the integration of teaching and technology. In addition to using traditional research methods to analyze learners' motivation, interaction patterns, the use of learning strategies and the roles of learners in the network environment, learning analysis and data mining techniques can also be used, this paper analyzes the learning trajectory, learning behavior and learning ability of learners by using educational big data. Big data can help to collect and analyze a large number of learner corpus, and make a visual analysis of the learning process in order to understand the complexity and dynamics of foreign language learning, and bring enlightenment to the design of foreign language learning module in the information environment, teachers and help learners to carry out personalized, autonomous learning.

Foreign language teaching in the context of information technology is a complex and dynamic system. All the aspects mentioned above are interrelated and inseparable. The analysis of concrete teaching cases at the micro level and the construction of the theory at the macro level are beneficial to a more complete understanding of the system and a deeper integration of information technology and foreign language teaching.

Oral English is one of the main difficulties that need to be overcome in English teaching at present. Students' oral English ability is one of the professional skills that will affect their future work, by means of group conversation, we can increase the chances of oral practice. There are many scene dialogues in English textbooks, which can make students act different characters for oral practice. In addition, through modern network resources, enriching more teaching content, such as playing foreign classic video clips, such as Pride and Prejudice, such films have strong appeal to students, can design dubbing links, let students practice oral English according to different situations and different difficulty levels, improve students' oral expression ability and self-confidence, and make oral English reach a professional level.

Facing the problem of too little communication between teachers and students in the classroom, first of all, we need to design reasonable teaching content, increase the interaction between students through group study, and set questions and answers for the knowledge points in English textbooks, students in the same group to actively exchange the answers to the questions, at the same time, the teacher and students in the discussion process, also let the teacher-student relationship closer, more harmonious.

In addition to interactive time in the classroom, teacher-student interaction is extended beyond the classroom through modern information technology, making full use of spare time, setting up some online activities, assigning learning tasks, and increasing opportunities for student interaction, improve students' interest in English and cultivate students' good habit of learning English independently. Teachers can make use of modern information technology to collect more English knowledge, expand the scope of knowledge learning, and share learning resources with students on the Internet. Through such teaching methods, students' comprehensive English ability can be developed, it can also broaden students' horizons and develop into English talents urgently needed by the society.

5 Conclusion

Under the background of the continuous development of modern information technology, the reform of English education is constantly being optimized and innovated. At present, with the support of modern information technology, English courses have been carried out in various teaching modes, mainly using modern equipment, such as multi-media, slides and other playing equipment, to make the English classroom more vivid, more intuitive, the past teaching in the abstract, difficult to understand teaching knowledge, using modern information equipment to display, in order to improve students' interest in learning, more can play learning thinking and enthusiasm, gradually achieve "Happy Learning", improve the quality of classroom teaching.

References

1. Jiang Fengchun, Si Bingyue. Multi-dimensional research on the autonomous teaching ability of college English teachers driven by information technology [J]. Foreign Language Research, 2017, (6): 3-59+112.
2. Liu Kun. The construction of English ecological classroom in the era of "Internet +" [J]. Teaching and Management, 2017, (36): 98-100.
3. Yu Yan. On the current situation and countermeasures of information technology in English teaching [J]. Education Modernization, 2017, (44): 286-287.
4. Jiang Lingling, Zhan Jianjia. Reflections on Integrating English Teaching and Internet Technology [J]. Education Review, 2018, (12): 165.
5. Zhu Ronghua. The supply-side structural reform of college English teaching from the perspective of information technology [J]. Modern Education Science, 2018, (12):

79-83.

The Research of Network Ideology Education in Hainan Higher Vocational Colleges under the New Media Vision

Yulan Ning, *Qiyue* Pang

Hainan Vocational College of Political Science and Law, Hainan, Haikou 571100, China

Abstract. In the view of new media, when Hainan higher vocational colleges carry out the education of network ideology, they should actively hold high the banner of the main melody, carry out in-depth the great action of Hainan civilization, and strengthen the mainstream ideological public opinion by consolidating, we should carry forward the main theme, dare to respond to the wrong trend of thought and the public opinion that confuses right and wrong, study the changes of the public opinion on the Internet, improve the control ability of the public opinion, seize the new highland of the propaganda thought, and improve the quality of the education of the network ideology in an all-round way. Based on the new media field of vision, this paper makes a systematic study on the network ideology education in Hainan higher vocational colleges.
Keywords: New Media, Hainan, Higher vocational colleges, Network ideological education.

As the main body of ideological and political education, the educators, especially the front-line ideological and political educators, should master all kinds of new media technology as soon as possible, adapt to the new media environment, and improve their new media literacy, meet the new challenge to the actual effect of ideological and political education of college students in the new media environment.

1 Under the new media visual field Hainan Higher Vocational College Network ideology education background

1.1 Ideological education of college students faces greater challenges

Before the emergence of Internet media, when colleges and universities carried out ideological education for students, they used such methods as teaching materials, newspapers and periodicals, bulletin boards, talks, etc. . These publicity and educational contents, is the result of careful screening. However, since the emergence of the network media, it is difficult to filter its content effectively when information is disseminated because of its high speed, wide content and multi-domain coverage, the content of information transmitted in this context of spatial freedom is difficult to distinguish between good and bad. Therefore, the network media by a variety of ideology and values of the pursuit. As for college students, in the stage of receiving higher education, it is the key stage for their moral concepts to mature and solidify, and their ability to distinguish corrupt values is not strong, if they are infected by negative information for a long time, they will distort their values and aggravate their inner contradictions, thus gradually deviating from the track of correct values, even the change of ideal and belief, the loss of values and the weakening of legal consciousness, all these problems have posed great challenges to the ideological education of college students.

1.2 The discourse power of ideological education is gradually weakened

The liberal democracy advocated by the Internet media will not be short of enabling all kinds of negative information to be spread. These ideologies, which represent different interests and invert the concept of right and wrong, will also be greatly exaggerated by means of the online media, so that the original correct ideology, in the expression of the right to speak, its influence and dominant position has been constantly weakened. At the same time, the privacy and virtuality of the network media will also make some negative content, take this opportunity to be spread, thus creating great convenience for its counter-mainstream thought and negative energy spread.

1.3 Ideological security and stability are undermined

The safe and stable development of ideology is not only the core content of the security and stability of colleges and universities and the construction of campus culture, but also the key content of the national security and stability. Therefore, ideological security and stability on campus stability, social stability and national ideological security, has a decisive role. After the emergence of the internet media, some countries in Europe and the United States used the internet media to demonize China, pass on negative information affecting national security to young Chinese students, and even fabricate some false news, in this way, some ideas that go against the mainstream values are infiltrated into the masses of students, and the use of Internet media to spread rumors and create incidents, so as to arouse the outburst of racial and religious emotions among some students, delusion causes the collapse of ideological education in colleges and universities, and then affects the safe and stable development of ideological education. Because of the openness and anonymity of the new media environment, some college students who are in the period of forming their world outlook, outlook on life and values, it is easy to be deceived by lots of information in the new media environment, which poses a new challenge to the ideological work in colleges and universities.

2 The characteristics of network ideology education in Hainan higher vocational colleges under the new media visual field

Ideological security education in higher vocational colleges has always been a matter of concern, because colleges and universities are important places for training future talents. With the rapid development of digital technology, new media technology has also emerged. Different from the traditional press, radio, television and other media, it is characterized by the use of network technology to provide teachers and students with information and services. The inherent mode of ideological education in higher vocational colleges changes with the development of new media technology. At the same time, the characteristics of new media are a double-edged sword to the security of network ideology in higher vocational colleges. Compared with traditional media, new media has the characteristics of convenience, pluralism, speediness and openness, it provides a platform for the ideological collision and cultural exchange between teachers and students. However, the services provided by new media technology have an impact on the ideological security of higher vocational colleges. Its interactive, virtual and open nature allows students to access these fragmented information, changes their perceptions and perceptions, and their thinking jumps.

Compared with the traditional media, the new media more in the form of digital presentation to you, it has a wide range of channels, easy to promote, rich in form and so on. For other groups, students in higher vocational colleges pay more attention to the fate of the country and the form of the country. Some students are prone to extreme and emotional words when they make their own statements, due to the existence of forums, microblogs and circles of friends, inappropriate remarks will soon spread to the whole country, causing adverse effects. As the students of higher vocational colleges come from all over the country, once there are public opinion events, it will form a larger social communication. Therefore, it is necessary to have the ability to judge right and wrong, especially when it concerns the interests of multiple countries and heroic deeds can not be misinterpreted.

With the popularization of the internet, students acquire knowledge in more and more diverse ways. In the Digital Age, book knowledge can be acquired through the Internet. On the network many high-quality course resources, unifies the student's actual to carry on the screening. Educators can use wechat, QQ and other apps to guide students' Learning Outcomes. The timely application of new media makes the mobile teaching run through the daily life without the restriction of time and place. The network new media has broken the traditional indoctrination education pattern, it may satisfy the teacher and the student communication need. The teacher is the leader, first of all must create the question situation, the student's independent study ability may obtain the training. Through real-time interaction and message board message content, to understand the students' real psychological needs. Students are more free to express their psychological thoughts and can communicate with others at any time. The teachers then gave the correct instructions.

Compared with the traditional ideological work, the network ideology has the characteristics of guidance and concealment. All sorts of bad emotions hide in the objective rational content, mislead students in the network public opinion, and finally lead to the infiltration of non-mainstream ideology. In particular, some Internet chat groups in a closed communication environment, talk about the topic will break the legal and moral bottom line, and then people's cognition and thinking will be affected.

3 The strategy of network ideology education in Hainan higher vocational colleges under the new media visual field

In the new media era, the network has become the first public opinion field and the main battlefield of public opinion struggle. Colleges and universities should also attach importance to this position. In addition to running their official websites, microblogs and wechat, they should also pay attention to hot issues on the Internet, organize teachers and students to discuss or hold lectures, in view of the problems among teachers and students, we should guide them to view the problems with Marxist standpoint, viewpoint and method, and encourage them to spread positive energy on the internet, so as to consolidate and strengthen the mainstream ideological and public opinion. Under the background of increasingly fierce ideological struggle, party schools in colleges and universities should dare to take on the responsibility, to carry out targeted research on the training of teachers, staff and students to join the party and on the theoretical and practical problems of party building and ideological and political work in colleges and universities, in particular, we should vigorously criticize the erroneous ideological trends of "universal values", "constitutional democracy", "historical nihilism" and "neoliberalism" among the activists who have joined the party.

In order to improve the new media literacy of the ideological workers in colleges and universities, it is necessary to pay more attention to the new media literacy of the ideological workers in colleges and universities, to regularly launch new media technology competitions for ideological workers and organize relevant training and study of new media technology, for example, the application of commonly used new media teaching software, the making and use of micro-lessons, the development and design of MOOCS, the theory and practice of the deep integration of information technology and education, etc. , make full use of new media resources to improve themselves, including the formation of advanced new media concepts, the use of new media technology, effective use of new media resources and reasonable criticism of the media. Ideological workers in colleges and universities should try their best to become experts in using new means and methods of modern media. Young people and old comrades should also learn from each other's strengths. To keep "charging" and keep learning, we should not only have a deep understanding of the party's theory, but also be highly sensitive to hot issues, and form a more acute political observation and discrimination, to elaborate the theory accurately and authoritatively, to constantly improve the ability and level; and to require oneself to study the law of the development of the network and the use of new media with the highest standards, improve oneself to be competent for ideological work under the condition of network.

In order to ensure the correct direction of Marxist education, we must improve the political and moral quality of teachers and take responsibility for it. As the object of ideological education for college students, some college students will, because of the existence of various factors, show that the level of media literacy is not enough, thus making their use of network media, lack of effective identification of bad information, which makes the role of the network media in ideological education difficult to play. In order to improve the media literacy of contemporary college students, we must start from the classroom, and the teachers should raise the level of media literacy of contemporary college students in an all-round way through the detailed teaching of various basic principles.

Under the new situation, the teachers of ideological identity education in higher vocational colleges should strengthen the awareness of big data application. The integration of big data into ideological education in higher vocational colleges can effectively improve the quality of teaching and improve students' initiative in autonomous learning. School administrators should also strengthen the awareness of big data, whether in the daily school management, or the ideological education of college students, the importance of the application of big data is becoming more and more obvious. Teachers in higher vocational colleges should have the consciousness of mining big data, so as to make better use of big data technology in higher vocational colleges to establish the mechanism of mainstream ideological education. Compared with traditional teaching methods, big data breaks the traditional teaching methods and ideas. Teachers should strengthen the consciousness and thinking of big data in their educational ideas and teaching contents. With the rapid development of big data, the mainstream ideology education in higher vocational colleges can achieve better teaching results. Changing the traditional teaching method and using big data information technology to carry out online teaching can not only expand the educational resources of the mainstream ideology in the teaching classroom, but also effectively improve the teaching quality of teachers, as well as in the new situation to improve the vocational college students' values and mainstream ideology of training.

Before the arrival of big data in higher vocational colleges, the education of Students' mainstream ideological identity in higher vocational colleges was mainly propagandized by means of TV, newspapers and bulletin boards, the way of publicity in higher vocational colleges has also gradually changed. Instead of continuing to use traditional propaganda, new media such as wechat and Weibo are used according to the development of modern science and Technology and information technology, in order to facilitate the dissemination of information more extensive and in place. As a concentrated group in network media, higher vocational college students have a better understanding of the mainstream ideological identity education of higher vocational college students. The main feature of the propaganda is to highlight the importance of mainstream ideological identity education in a comprehensive way. Ideological identity education in higher vocational colleges can be carried out online education work, will have excellent theoretical results of the network courses published to the channel for students to watch, so as to improve ideological education work. It is necessary to strengthen the communication function in the online channel so as to make the students better express their suggestions and ideas, and thus help the teachers to understand the dynamics of the ideological identity education of the students in higher vocational colleges and optimize it.

4 Conclusion

With the rapid development of network new media technology, the study of ideological security education can not only promote the development of ideological education in colleges and universities, but also indirectly help college students to establish a correct awareness of network security. The young college students are in the important stage of the maturity of the socialist ideology. To set up the correct ideology will help to set up the correct ideals and beliefs. The network information is widespread and many kinds of information mix together, if lacks the rational thinking to the thing judgment, easy to produce the emotion fluctuation and the choice confusion, will cause the student to have the psychological question. Correctly guiding and improving the students' judgment ability in the face of the diversified network information is conducive to building a good campus atmosphere and improving students' mental health.

5 Acknowledgements

2020 college-level Project of Hainan Vocational College of politics and law, "research on the path of strengthening mainstream ideological education in Hainan higher vocational colleges under the background of the construction of free trade port"(subject number: KYYB2002)

References

1. Qin Yanping. Research on Internet Ideological Discourse Power in my country under the Background of Big Data [J]. Journal of News Research, 2017, (23): 28-29.
2. Song Xiaoqin. Realistic challenges and coping strategies of network ideology construction [J]. Gansu Science and Technology, 2017, (21): 4-6.
3. Shi Chao. The Construction of Internet Ideological Discourse Power [J]. Young Reporters, 2017, (26): 1-12.
4. Liu Haofan. Analysis of the importance of strengthening the construction of network ideology [J]. Popular Literature and Art, 2017, (13): 270.
5. Zhou Zhe. Challenges and countermeasures faced by Internet ideology in Guangdong universities [J]. Journal of Guangdong University of Foreign Studies, 2018, (6):

107-111.

A Study on the Application of Computer-assisted Translation in Henan Museum Translation

Yanan Sun

Foreign Language School, Zhengzhou University of Science and Technology, Henan, Zhengzhou, 450064, China

Abstract. Museum is the Treasure House of human history and culture, which bears the responsibility of inheriting history and spreading culture. The Henan Museum is also attractive because of its rich and profound traditional culture. In order to make the world fully understand Henan culture, it is necessary to actively explore scientific and reasonable English translation strategies based on Henan Museum text features and incorporating Trans-cultural diffusion features. In the context of intelligent translation scenarios, the advantages of computer-aided translation are becoming more and more obvious, and the full play of Computer-assisted translation will provide effective support for improving the quality of Henan Museum translation.
Keywords: Computer-assisted translation; Henan Museum; Literature Translation; applied research.

As a branch of translation, museum document translation plays a bridging role in the process of connecting Chinese and Western cultures and expanding China's foreign exchange. Therefore, it is of great practical significance to improve the quality of museum translation and strengthen the communication with the western museum circles. The Henan Museum is a microcosm of Central Plain's cultural history. With the advent of the information age, the demand for translation is increasing day by day, and supported by Computer-assisted translation and MT software, it will provide effective support for the English translation strategy of innovative Henan Museum.

1 Computer-assisted translation in the English translation of Henan Museum

The English translation of Henan Museum documents is highly professional and practical, and its translation should pay attention to information translation, accurate content and natural expression, the report adopts the model of tmxall + Google Translate + post-edit. The term base and memory base are established by using computer-aided platform, and the matching degree between the term base and the target text is analyzed. By comparison, it is found that the matching degree of terms and proper nouns is high, but the matching degree of sentences is low, which is mainly due to the lack of relevant corpus and low reference value. In addition, machine translation also has some limitations, the machine translation to simple sentences, sentence components complete processing effect is better, but for some non-subject sentences, passive sentence processing is not satisfactory.

Computer-assisted translation is a combination of translation memory and machine translation and human proofreading, a set of human computer interaction processes that are useful for non literary text translation. Computer-assisted translation. Generally speaking, Computer-assisted translation can be divided into two categories, the first is the general purpose software and hardware, which includes the more commonly used word processing software, E book software and some search engines, etc. , some of the hardware, such as erasable optical drives, digital cameras, and some scanning equipment, among which OCR plays an important role in the processing of language files before translation. Currently, most fan Yuyuan files are in multiple formats, especially those in image format, but Computer-assisted translation's software is currently unable to accurately identify and edit text between images, this can only be done with the help of OCR software technology, which is one of the most important tools for translators to deal with pre-translated source files. In the process of translation, some simple mechanical but repetitive work can be transferred to the computer, and better assistant translation technicians to translate, at the same time, it can help the work item to carry on certain processing and the analysis, can help the translator to set up the translation division of labor clearly and the effective conformity mechanism, realizes the more fluent translation work process.

The impact of Computer-assisted translation on translators. In general, Computer-assisted translation is concerned with the clever use of computer software to improve the efficiency of human translation, thus ensuring the efficiency and quality of further human translation, at the heart of this technique is the memory of the translation. The memory of translation can record and store the relevant content on a certain basis, which can significantly improve the efficiency of subsequent translation tasks, and can greatly help the translators to alleviate the pressure of work, to obtain a large amount of useful information in a short period of time so that it can carry out more translation tasks. The structure of the Computer-assisted translation system has certain advantages, such as improving the relative efficiency of translation and reducing unnecessary duplication of work for translators At the same time, it can ensure the consistency of the meaning between the translated text and the source text. In the whole translation, one source document can avoid the inconsistency of the translated language effectively through the cooperation of many people. At this point, the database of applied terms can ensure the consistency of the translated text and the source text, thus ensuring the quality of translation and the use of system structure can help the translator to improve the translation efficiency greatly, therefore, the quality of the whole translation can be guaranteed, and the translation process can be further speeded up by reducing the cost of translation and memorizing the data in the database.

2 The application of Computer-assisted translation in Henan Museum

The English translation of the Chinese version of the museum is a good bridge to introduce our country's long history and culture to friends around the world. But in the original language, the introduction of cultural relics is usually in accordance with the customs of some dynasties and proper nouns, for people with relevant background, these are easy to understand; However, for the target language readers who know little about Chinese culture and history, relevant background information should be included in the process of translation.

2.1 System architecture

As far as translation services are concerned, the Computer-assisted translation supports an online BS mode of operation that prevents redundancy when different translators recycle their libraries and reduces rework. Based on the result of automatic translation, the system optimizes and Perfects the result of automatic translation through the information of the intermediate result of machine translation decoder, Memory Bank, dictionary, etc. , at the same time, it provides multiple interactive modes for manual editing, which can realize the organic combination of machine translation and manual translation, optimize the user experience, and guarantee the translation efficiency and overall quality. The input-output function module transmits the original text to the system translation module, and then outputs the results based on the translation module, a balanced bilingual parallel corpus, in which the data are linked to the translation module, which receives the original text based on the input and output module, this method is based on the maximum reverse matching method and the corpus module, in which the sentences are matched mainly by the elicitation words to provide the most suitable translated sentences. The Computer-assisted translation's translation process, which automates the translation and generation of intermediate result data through machine translation of source language oriented documents, and decodes intermediate results through machine translation, based on the auxiliary translation information, the automatic translation results can be corrected by means of auxiliary translation adjustment, thus the target language documents can be generated.

2.2 Translation strategy design

In terms of Computer-assisted translation, the translation strategy is oriented to the actual needs of the translation users, and the translation results are produced through the system human-machine interface, and the translation corpus is used as the carrier to help the users complete the translation task. In the Computer-assisted translation system design, the translation strategy corpus can support a large number of bilingual source language sentence auxiliary balance alignment, support various file format conversion and word count statistical analysis, it has a functional corpus alignment device, a search function, which can query words and sentences in the translation memory and provide translation suggestions, and a sentence segment screening function, which can select the required sentence segments between the original text and the translated text; Support translation management function, easy for translators to achieve multi-document management, managers can distribute files through the management side, view the translation process, export translation results.

3 The application of Computer-assisted translation in Henan Museum translation

There are still many problems in word order and word selection for the translated documents generated by machine translation technology, but their readability is relatively high. After the implementation of the online machine translation engine, the word order of the translated documents has reached a level similar to that of the artificial language to some extent, and after the preliminary translation of reference materials to screen key information, the translation engine interface, in terms of daily interpretation, there is a trend to replace junior translators.

3.1 Understanding language differences and enhancing cultural sensitivity

Both English and Chinese belong to the indo-European language family and the Sino-Tibetan language family respectively, this paper analyzes the causes of cultural heterogeneity in English and Chinese from four aspects: cultural origin, language structure, thinking mode and values. Chinese sentence structure, often start from the verb, according to the action or semantic relevance of the arrangement of verbs, so Chinese is a dynamic language. The process of C-E translation is the process of transforming dynamic Chinese into static English. In this process, we first need to clarify the primary and secondary relations of Chinese verbs. Only the main verbs can be regarded as predicates in English sentences, the rest is accomplished by non-finite verb, prepositional phrases, and clauses. From the traditional point of view of rhetoric, this paper puts forward that Chinese is "fastidious about neat words and sentences" and "likes rhetoric and rhetoric", while Western discourse respects objective facts, does not like strong words and color leakage. A comprehensive comparison of languages will help us to establish the awareness of cultural differences and try to enter into the cultural context of information recipients, so as to seek more effective methods of cross-cultural communication.

3.2 Pay attention to the construction of the translation team and optimize the translation workflow

As translation is not only a transfer between languages, but also a re-creation of basic information, the translator's quality, working time and energy are required. Museums should avoid the marginalization of translators and the ambiguity of their roles, and provide them with adequate time guarantee, supplementary materials and personnel support, in this way, the translators in the embassy can better fulfill their duties and responsibilities. At the same time, as the library-school cooperation continues to advance, we should make good use of the present opportunity, quote the "external brain", properly recruit foreign professional translation scholars, foreigners, English majors to join the translation team, actively seek higher level of professional guidance to ensure the quality of translation. In order to guarantee the accuracy of the translation work, the translator should join the curatorial team as soon as possible and work with the curators while translating. This will be a process of mutual standardization between Chinese and English texts, the Chinese team provides the translators with accurate information interpretation and in-kind reference, while the translators are guided by the principle of "promoting cultural identity and achieving effective cross-cultural communication", remind and advise the team to increase the information background knowledge which may explain the explanation to the foreign audience. The cooperation of the team maximizes the efficiency and realizes the effective communication and standardized operation between the two sides.

3.3 Innovative dialogue mechanisms among civilizations

Museums should have the courage to undertake the mission of spreading the regional culture and the excellent traditional culture of the Chinese nation. In today's increasingly frequent international exchanges, seize the opportunity to explore their own development path, the development vision of municipal museums can not only be confined to domestic, but should have a broader international perspective, through the introduction and introduction of international temporary exhibitions, and the construction of exchanges, dialogues and exchanges with different civilizations, to further promote the international community's cultural level of equality, respect and inclusiveness of the cultural exchange model, strive for the "discourse right" of Chinese excellent culture in the world civilization.

In general, when translating a text, the translator has certain translation standards and principles, such as: Faithfulness and accuracy of the original text, fluency and truthfulness of the language, accuracy of the style of the translation and so on. Of course, these are also suitable for the English translation of the museum text; because of the particularity of the museum text, it contains many cultural relics with a long history and culture, and its own particular principles and standards. It is especially important for the translator to be faithful when translating, especially when translating the translation with special attributes such as museum relics and proper nouns. In the translation of the original, we must uphold the principle of fidelity.

4 Conclusion

The Chinese-English translation of museum texts is a special style which can introduce its history and culture to the outside world. The Chinese-English translation of the museum text belongs to the translation of applied style. The Chinese-English translation of cultural relics is especially special and difficult because of its particularity and long history, as well as its own national history and cultural imprint. This requires the translator to have a good command of the written language, to be familiar with the two cultures, and to know the purpose and standard of translation. Although machine translation has not directly hindered the core work of interpretation and translation, it has the potential to upend the translation ecosystem as it matures into what is known as disruptive innovation.

References

1. Pang Shuangzi. Bibliometric Analysis of International Corpus Translation Research in the Past Ten Years (2008-2018) [J]. Foreign Languages and Foreign Language Teaching, 2019(6): 129-140+149.
2. Zhang Yinan, Li Yanhong. Characteristics and principles of translation of museum texts: Taking Hubei Provincial Museum as an example [J]. Journal of Hubei University of Technology, 2019(3): 74-77.
3. Yan Xin, Chen Ruizhe, Zhang Jing. The development status and trend of translation technology cloud platform [J]. China Science and Technology Translation, 2019 (1): 22-25.
4. Long Minghui. Keeping pace with the times and innovating translation: On the re-translation of Chinese classics in the digital age [J]. Foreign Languages (Journal of Shanghai International Studies University), 2020 (2): 121-128.
5. Zhou Xinghua, Wang Chuanying. Application and evaluation of artificial intelligence technology in computer-aided translation software [J]. Chinese Translation, 2020(5): 121-129.

Artificial Intelligence Platform Construction and Integration Based on Multi-sensor Fusion

Hongsheng Wu, *Junwang* Hu*

School of Professional Technology and Education, Guangxi Science & Technology Normal University, Laibin, Guangxi, 546100
*Corresponding author: Junwang Hu

Abstract. Sensor is an important source of data transmission, through the integration of mechanical radar, inertial measurement unit and other sensors, building an artificial intelligence platform, will effectively save data information. By taking advantage of sensor fusion, the data of gyroscope and accelerometer are transferred to the background, and then the time stamp of artificial intelligence platform is aligned to get useful and usable information. By combining the mechanical radar and millimeter wave radar, an artificial intelligence platform is built, which makes full use of the synergetic advantage of the related sensors, captures the data information independently and saves it according to the time stamp. This paper focuses on the construction and integration of artificial intelligence platform based on multi-sensor fusion.
Keywords: Multi-Sensor fusion; artificial intelligence; platform building; integration research.

The development of artificial intelligence depends on the combination of algorithms, data and computing power to achieve better results. At this stage, computing power needs a breakthrough in hardware, the progress of algorithm needs more algorithm engineers' efforts to have a breakthrough. Whether can use the data efficiently is each big company's disparity, the artificial intelligence data center is solves this question the best solution without doubt. Therefore, take the lead in building, continuous optimization, platform intelligence in the direction of a big step forward.

1 Artificial intelligence platform based on multi-sensor fusion

In modern control systems, sensors are located at the interface between the object under test and the system under test, and constitute the main "window" for system information input, it provides the original information necessary for the control, processing, decision-making and execution of the system, and directly influences and decides the function of the system. Multi-sensor integration and fusion is a research hotspot in the field of intelligent machine and system at present. Based on the types and types of sensors selected, their physical connection mode is designed: most of them are connected by Ethernet, a few are connected via the CAN bus. And on the basis of the SDK of each sensor, the control of reading, writing, circulation, pause and other functions of each sensor is realized, a man-machine user interface (GUI) was developed to allow users to choose which sensors to start to collect data. The artificial intelligence platform can be used for laboratory personnel to push out to collect data, and later researchers can use algorithms to study the collected data. The artificial intelligence platform provides the tool kit for the user to build the intelligent application. These platforms combine intelligent decision algorithms with data to enable developers to create business solutions. Some platforms provide pre-built algorithms and simple workflows, with drag-and-drop modeling and visual interfaces to easily connect the necessary data to the final solution, other platforms require more development and coding knowledge. In addition to other machine learning functions, these algorithms can include image recognition, natural language processing, speech recognition, recommendation systems, and predictive analytics.

The AI platform is an infrastructure for building large-scale intelligent services, providing step-by-step building and life-cycle management of required algorithmic models, so that we can sink our business into algorithm model to achieve reuse, composition innovation, scale building intelligent services and the role of business empowerment. Each application continuously produces data, each business module data collection, after a unified cleaning, classification, error correction, labeling, definition, granulation and indexing, data center. Then according to various algorithms and machine learning, thus forming artificial intelligence platform.

With the development of artificial intelligence, the Internet platform will be more intelligent and informative. The emergence of multi-sensor integration and fusion technology is precisely to solve these problems faced by a single sensor, multiple sensors can not only be described the same environment characteristics of multiple redundant information, but also, can describe different environmental characteristics, for example, a mobile robot with multiple sensors, it can get the distance information of an object through stereo vision and ultrasonic sensors, respectively, while, you can also get information about the shape of the object visually.

The use of multiple sensors can also parallelize the process of data acquisition and processing, thus improving the performance of the whole system. In addition, other technological developments. Especially, the development of computer technology and sensor technology lays a foundation for the development of multi-sensor integration and fusion technology in both hardware and software. Compared with single sensor system, the integration and fusion of multiple sensors can not only obtain more comprehensive and accurate information, but also reduce the cost and time. To sum up, the characteristics of multi-sensor integration and fusion can be simply summarized as redundancy, complementarity, timeliness and low cost.

2 The mode of artificial intelligence platform construction and integration based on multi-sensor fusion

Artificial Intelligence and big data platform is one of the main research topics in the industry. The organic fusion of artificial intelligence and big data platform will make the big data platform more intelligent. Based on multi-sensor fusion, an artificial intelligence platform will be built, it is of great practical significance to widen the application field. The integrated pattern and framework mainly studies how to build a general information processing pattern for multi-sensor fusion, that is, the fusion structure, at present, the commonly used methods are layered phase plane pattern, neural network, logic sensor and object-oriented design. In the layered phase-plane model, information processing is divided into four stages according to the time and scope of perception, namely remote stage, approaching stage, contact stage and Operation Stage. In a neural network, trained neurons are used to represent sensory information, and through associative memory, neurons are assembled in complex combinations to respond to different sensory stimuli, the physical sensor is abstracted as a logic sensor, which provides a unified framework for multi-sensor integration. It can adopt the network structure as the structure of information processing. In object-oriented design, sensors can be expressed as objects, and objects can communicate with each other by messages, the sensor's processing can be invoked with the message. Neural network method does not pay attention to the details of information processing, but only cares about the mapping relationship between input and output. It has a good application prospect in some fields such as object recognition. Logic sensor and object-oriented design methods are more convenient for the design and implementation of multi-sensor system. At present, there are many systems using these two methods, the last two methods pay more attention to the details of information processing, both using layered structure, which provides a flexible framework for multi-sensor integration and fusion.

The content of fusion method research is the algorithm related to data fusion. The essence of multi-sensor fusion is the processing of multi-source uncertain information. It's a complicated process. As mentioned before, information in the system from the bottom-up process, the presentation of information in the form of constant change. In addition, the uncertainty of information can be in the form of random, fuzzy and prior information, or in the form of no prior information. There are different processing methods for different information representation forms. This paper introduces some typical fusion methods, such as weighted average, Kalman filter, Bayesian estimation with consistent sensors, multibayes, statistical decision theory, evidential reasoning, fuzzy logic and production rules.

The weighted average method is the simplest and most intuitionistic fusion method. It is suited for dynamic environments. It performs a weighted average processing of a set of redundant raw sensor data as a result of final fusion; Kalman filter is widely used in multisensor fusion systems. It is suitable for processing dynamic, low-level, redundant data. Its real-time performance is good; the specific step of adopting the Bayesian estimation method of the uniform sensor is to first remove the sensor information that may be wrong, then the redundant information provided by the remaining "consistent sensor" is fused. It is suitable for static environments; in a multibayes approach, all sensors are considered as a group of multiple decision makers who must determine the group's consistent results.

Since most of the algorithms are implemented in Python, we'll wrap the Python implementation in an HTTP service, and we'll start with the flask framework in Python. The clustering implementation of the algorithm, we use NGINX to do reverse proxy forwarding, an nginx cluster, can control many different algorithm clusters (such as each different type of algorithm to build a cluster) , with the sidecar component, the algorithm service is also registered in spring cloud as a service, and then called the algorithm service as sping cloud internal service. All the hardware (mechanical radar, millimeter wave radar, Tripod head, dual-beam camera, IMU, industrial camera, mainframe Ubuntu, switch) needs to be connected together, the mechanical radar, dual optical cameras, IMU, industrial cameras through the switch with the host Ubuntu Connection. Millimeter-wave radar, pan-tilt through USB2CAN and host Ubuntu connection.

3 Artificial Intelligence platform construction and integrated design based on multi-sensor fusion

The research content of control structure is how to effectively control the process of multi-sensor integration and fusion, according to the characteristics of different applications, there are three methods: sensor and control hierarchical structure, distributed blackboard and adaptive learning. The hierarchical structure of sensing and control is composed of three parts: sensing process, task decomposition process and the world model connecting them, through the cooperation among these three parts, the hierarchical control is realized, a hierarchical structure is more suitable for handling complex tasks in a distributed blackboard structure, which allows communication between subsystems and effective distributed control through the information center. The structure is especially suitable for real-time applications with certain requirements.

In adaptive learning, the system can find the appropriate control signal according to the output of the sensor. This ability is to be acquired through training. This structure is suitable for application control in dynamic environment, but there are still some problems in the learning method itself, such as the generalization ability of neural networks and whether it can be online learning and so on. Sensor selection is part of multi-sensor integration, which enables a multi-sensor system to select the most appropriate combination of available sensors. Pre-selection method and real-time selection method are two basic methods at present. Pre-selection method is an optimal method for sensor design. It is based on the specific application and optimal design criteria, prior to the optimal configuration of sensors. Compared with the pre-selection method, the real-time selection method is used to configure the sensors in the process of system operation. According to the change of the system, real-time configuration can be made to achieve local optimization. The applications of the two methods are different. The former is suitable for static environment, while the latter is suitable for dynamic environment.

Given the highly scalable nature of the system and the many other services that may be added to the platform in the future, and the potential for smart innovation, when it comes to web framework selection, the extensibility of the system must be prepared in advance. Since the service was not implemented initially, the focus was on the iwogh-web module. All requests are still subject to reason control using the ZUUL module in springcloud. Eureka's in charge of the service registry. The Mysql database that the database uses, because the data is not big at the beginning, the Mysql of concern type can satisfy the demand, do not need to consider the solution of big data in the beginning.

In order to control all sensors on a GUI, the project uses Python tkinter to develop the interface and manages the executables of multiple sensors in a multi-process manner. Each sensor's data is named after a time stamp and stored as a file. But since these timestamps are from Ubuntu, the host, and not the actual sensor, they can cause errors and can be time stamped aligned. For example, if you need to know the absolute time of a mechanical Lidar, you need to provide a GPS module. When a large amount of data collected by each sensor is obtained, the data processing algorithm can play a corresponding role. Different from the traditional analytical database, DRUID provides real-time stream data analysis, and uses LSM-Tree structure to make Druid have high real-time writing performance At the same time, the visualization of real-time data in sub-second level is realized.

4 Conclusion

Because of the complexity of object and process, there is not a systematic method to solve all the problems in multi-sensor fusion. Although there are many fusion methods, each method has its own scope of application. It is obviously impossible to solve the problem in a single way. The ideal solution is the integration of multiple fusion methods. Therefore, how to choose specific fusion methods according to specific problems and how to organize these methods are worth studying.

References

1. Gao Yun; Lu Hu; He Leinan. Plug-and-play multi-sensor fusion method for smart terminals [J]. Chinese Journal of Detection and Control, 2019(5): 83-89.

2. Tang Hao; Zhang Zilin; Zhou Bifeng; et al. Motion sequence optimization of the docking platform between sensor smart chips and array optical fibers [J]. Chinese Journal of Transportation Engineering, 2019(5): 53-63.
3. Tao Wuqiang; Wang Jie; Jiang Chao; et al. Multifunctional optical fiber sensor based on SMS structure [J]. Laser Journal, 2020(12): 13-17.
4. Liu Xiaokang; Li Jiahao; Peng Kai; et al. Theoretical model and error analysis of planar two-dimensional time grating displacement sensor [J]. Journal of Instrumentation, 2020 (12): 111-121.
5. Wang Rui; Chen Zhe; Yin Fuliang. A review of distributed acoustic sensor array calibration methods [J]. Chinese Journal of Electronics, 2021(12): 2468-2478.

Research on the Integration of Information Technology and Ideological and Political Course under the Background of Artificial Intelligence

Xinxiang Hua

Zhengzhou University of Science and Technology, Henan, Zhengzhou, 450064, China

Abstract. Under the artificial intelligence environment, the integration of information technology and curriculum teaching is deepening, which provides an important impetus for optimizing the classroom teaching form, integrating high-quality resources and promoting the three-dimensional upgrade of curriculum teaching. In order to realize the ideal value of ideological and political education, it is necessary to construct an intelligent ideological and political course based on the mature application of information technology, fully display the vitality of ideological and political education classroom, in order to achieve the ideal effect of ideological and political education. Through the application of artificial intelligence technology as the research background, systematically explore the integration of information technology and ideological and political courses path.

Keywords: Artificial Intelligence, Information Technology, ideological and political course, fusion research.

In the application of each educational technology, the collision and integration of knowledge and technology will be embodied in the whole process of design, development, utilization, management and evaluation. The application of artificial intelligence technology in ideological and political education can not be limited to the theory of ideological and political education and the concept of Zhao Kuo, we should examine how the benefits of intelligent education technology can be realized in all aspects of ideological and political teaching design, so as to embed the idea of intelligent age into ideological and political teaching.

1 The new changes of ideological and political course under the background of artificial intelligence

Artificial intelligence technology in ideological and political teaching methodology, showing a "thought and reality" Unity of opposites. On the one hand, this contradiction is reflected in the mismatch between the advanced concept of artificial intelligence technology and the outmoded construction of intelligent teaching concept. On the other hand, this is the common development opportunity of the potential market of wisdom education and the inflection point of ideological and political teaching reform in colleges and universities. In order to realize intelligent education, ideological and political education in colleges and universities has gradually changed to a style of accuracy, pluralism, autonomy and coordination, teaching decision-makers also recognize the possibility that artificial intelligence technology brings for the diversity development of teaching material system. In order to conform to such a complicated relationship between "thought and reality", the development of the methodology of ideological and political course in colleges and universities in the era of artificial intelligence must be a combination of differentiation and unification. It not only needs to keep the position of the basic theory of Marxism in the teaching material system, but also needs to adapt to the needs of the times and use new teaching techniques to subdivide the forms of knowledge expression.

After the unified content frame of ideological and political education policy, the differentiation trend of ideological and political education method development is the inevitable result of the intellectual development of ideological and political education. There are two main forces driving this divergence. The first is the external motive force, which originates from the persistent motive force brought by the realistic problems of social development and the need for ideological and political education in universities to fulfill its social service responsibility. With the advance of network construction, the contact point of information tends to be ubiquitous and the internalization of knowledge becomes more and more scattered. In order to help students analyze and solve real life problems better with the basic stand, viewpoint and method of Marxism, the content design of ideological and political theory also needs to get rid of the block-cramming teaching, subdivided teaching of small unit knowledge structure. The second is the endogenous motivation, which mainly comes from the desire of the ideological and political education subject itself for cross-domain knowledge production. In the era of globalization, due to the intensification of competition, the complexity of social problems has been gradually increased. These complex problems are not solved by a single knowledge structure of ideological and political science, the theory of ideological and political science needs multi-disciplinary and multi-domain thought, which also makes the process of multi-domain subdivision of ideological and political education speed up and expand.

The development of wisdom education also puts forward higher demand for the professional level of ideological and political teaching technology application. On the one hand, ideological and political education workers need to be familiar with the application of different intelligent education technology scene, in order to fit the teaching effect of different teaching links. On the other hand, the professional quality of ideological and political education workers is still the core of the application of intelligent education technology, ideological and political education is not the guarantee of deviation. Therefore, it is particularly important to make use of the adaptive characteristics of artificial intelligence technology to break through the barriers of ideological core and technology application of ideological and political educators. The advantage of adaptive feedback is that it can realize the automatic design of ideological and political education content by digital calculation, but we still need to guard against the separation of knowledge and value of ideological and political theory content. In order to avoid this situation, ideological and political education needs to emphasize the subjective status of the educators and the Educatees, and improve the construction of knowledge atlas with the help of core professional knowledge, adaptive feedback to realize the integrity of ideological and political education.

For the educators, the ideological and political teaching can take the core knowledge principle as the node, take the mainstream ideology wind as the boundary line, construct the dynamic knowledge atlas which has both the knowledge and the value. As for the educated, they can give full play to their subjectivity, internalize the structure of ideological and political knowledge and improve their ideological and political moral level. The two interact with each other to give full play to the guiding role of ideological and political education and to ensure the richness and integrity of the content of ideological and political education.

2 The challenges of the integration of information technology and ideological and political education in the context of artificial intelligence

With the development of science and Technology, artificial intelligence technology is in a process of rapid perfection and maturity. The development of artificial intelligence has brought about a brand-new educational and teaching idea, promoted the innovative thinking in the field of education, and provided a strong technical support for the educational reform in the new era. However, how to fully integrate AI into the field of education, especially into the ideological and political education, is still in the exploratory stage, AI education application is not yet fully mature, to a large extent, still need to continue to explore.

The key to the effective integration of education and AI technology is the effective integration of teachers and AI technology. Faced with the rapid development of artificial intelligence technology in the information age, ideological and political teachers in colleges and universities must quickly adapt to the requirements of this situation, be good at learning, and master the relevant knowledge in the fields of artificial intelligence and information technology, to continuously improve the ability of applying new technologies, and to embody and transform technological advantages into their own professional advantages and teaching abilities, so that artificial intelligence and other advanced information technology to improve their ideological and political education level of an important means. This form of ideological and political course of the university teachers of the overall quality of the higher requirements. However, most of the teachers in ideological and political courses are liberal arts graduates, and their proficiency in AI technology is insufficient, which makes the advantages of AI technology can not be fully and effectively demonstrated in the educational process.

From the point of view of the object of ideological and political education, artificial intelligence has limited ability to discriminate and process various kinds of information, such as not being able to fully identify and filter some poisonous and harmful information, these good and bad mixed information become a hidden danger in the ideological and political education. The high frequency of use of the Internet and new information technology tools by the entire social group at present is among the young people whose values are at a critical stage of formation and development. Under the artificial intelligence technology background, the internet information presents the characteristics of magnanimity, fast transmission, barrier-free, and so on, but the youth group sometimes lacks the correct value judgment, they can not screen out the useful information and filter out the harmful information, which makes them easy to be misled by the bad information and affect the positive guidance of ideological and political education in schools.

The goal of the ideological and political education in colleges and universities is to guide the ideological and political aspects and to educate the core socialist values of college students in accordance with the requirements of social development and the laws and practical needs of the development of college students' spirit, ideology and cognition, to form a correct view of the world, values and outlook on life. In the process of ideological and political education in colleges and universities, ideological problems account for the vast majority. Ideological and political teachers should be good at making students be educated and influenced in the positive emotional experience state through the imperceptible education method, and then generate a healthy and stable mental state to achieve the goal of ideological and political education. In this process, ideological and political teachers should flexibly use a variety of teaching methods and means, be good at grasping the best educational situation and opportunity, so that students can be fully edified. At present, there are still some gaps between AI and imitation of human teaching behavior, and these gaps are often the key and focus of ideological and political education. This kind of technical restriction is very obvious, if can not improve, will make the education effect big discount even in the form, then affects the ideological and political education goal realization.

3 The integration strategy of information technology and ideological and political course under the background of artificial intelligence

The application of information technology to teaching will have a double-sided effect, we should look at it dialectically and try our best to get rid of the negative influence of information technology on ideological and political teaching. No matter how the information technology develops, the teacher is always the leader of the classroom, and the students are guided by the study. Under the condition of information technology, in the cloud classroom, massive open online course and online-to-offline hybrid teaching process, first of all, ideological and political teachers should actively integrate various teaching resources before class to improve the control of the curriculum. In the era of network information, teachers should pay attention to the hot issues and focal points at home and abroad, and integrate them into classroom teaching, set up problems, guide students to discuss and criticize the wrong opinions at home and abroad, Under the environment of information technology, ideological and political teachers should change their ideas, keep pace with the times, learn and improve the ability of information technology application, and integrate it into classroom teaching. First, strengthen the training of information technology application ability, through the way of phased training, the first stage is the basic application stage, the second stage is the skilled application stage. Some older teachers who have little contact with information technology will be arranged as the first stage. Through training, they will master the basic skills and solve the problem of "how to use information technology" in ideological and political education. Young teachers and middle-aged teachers can be arranged into a second phase, which focuses on "what to do" to better integrate information technology into ideological and political education, training can take the form of "go out" and "come in". Secondly, teachers of ideological and political courses should also learn to rationally screen and choose all kinds of information technology, not blindly follow the integration of information technology into ideological and political teaching, and choose educational methods suitable for themselves, technical means adapted to educational content. At the same time, ideological and political teachers in colleges and universities should set a good example for students by setting a good example.

Although ideological and political course teachers in colleges and universities have a solid theoretical foundation, they are relatively deficient in information technology literacy. But the ideological and political course and the information technology deep fusion, the ideological and political course teacher's comprehensive accomplishment is very key. Therefore, colleges and universities should strengthen the ideological and political course teachers information technology training, constantly improve the ideological and political course teachers information literacy and modern education capacity. First of all, colleges and universities should organize ideological and political teachers to carry out information technology training so as to enable them to master information technology knowledge and related skills. At the same time, the innovative development of information technology will never stop, which requires colleges and universities to increase the study of information technology theory on the basis of imparting information skills when training ideological and political course teachers, to promote their awareness of the advantages of information-based teaching, so that they can consciously learn information technology in practice, to ensure that their information literacy level to meet the information technology and ideological and political education in-depth integration of the requirements of teaching. Secondly, ideological and political teachers should organize regular teaching experience exchange meetings, study excellent deep integration cases together, and constantly enhance their sense of integration and ideological and political information teaching ability.

Teaching structure is one of the basic elements of ideological and political education, which is composed of teachers, teaching materials, students and teaching media. First of all, from the perspective of ideological and political teachers, ideological and political teachers should change their teaching concepts, take ideological guidance as their main teaching goal, and make effective use of information technology to create teaching models that meet the actual needs of students, to provide a guarantee for carrying out targeted ideological and political education. Secondly, based on the perspective of students, students should actively participate in ideological and political class teaching design, give full play to their cognitive advantages of information technology, and make suggestions for ideological and political class teaching design, continuously improve the participation of students in the classroom, for the information technology and ideological and political education teaching in-depth integration escort. Finally, from the perspective of teaching materials and teaching media, teaching materials and teaching media should complement each other on the basis of integration, so as to make the teaching contents of ideological and political courses conform to students' reality, it is convenient to improve the actual effect and quality of ideological and political education. Teaching media is the embodiment of information technology. We should base on teaching materials and perfect the function of multimedia, so as to meet the needs of ideological and political education and realize the integration of information technology and ideological and political education.

4 Conclusion

In recent years, the extension of ideological and political education has been expanding, from knowledge-based education to wisdom-based education. The ideological and political education in the new period not only pursues the simultaneous coexistence of knowledge and values, but also realizes the simultaneous development of morality and wisdom, and restores the true essence of the ideological and political course of "truth, purity, reality and liveliness." Realize "change knowledge into wisdom". The thinking of wisdom education of wisdom education system is in line with the development demand of ideological and political education transformation. Wisdom education emphasizes a kind of educational idea and value pursuit, which can inspire people's wisdom and innovation ability, promote the actual effect of education and cultivate people with wisdom.

5 Acknowledgements

This work was supported by Key Research Projects on Humanities and Social Sciences in Henan Universities in 2022 under Grant 2022-ZZJH-336.

References

1. Qiang Biao. Research on the Transformation of Ideological and Political Education

Curriculum in Colleges and Universities Based on Essential Reduction Theory——Also on the Dilemma and Way Out of the Practice of Ideological and Political Courses [J]. Jiangsu Higher Education, 2017, (12): 53-55.

2. Zhang Baojun. Reflections and strategies on the teaching of ideological and political theory courses in colleges and universities from the perspective of "precise supply" [J]. Ideological and Theory Education Guide, 2017, (8): 99-102.
3. Xu Shuang. Analysis of the teaching mode reform of ideological and political courses in colleges and universities in the "Micro Era" -- Commentary on "Interaction Research on the Teaching Reform of Ideological and Political Theory Courses and Ideological and Political Education of College Students" [J]. China Education Journal, 2018, (3) :129.
4. Wang Xuejian; Shi Yan. The connotation, characteristics, difficulties and coping strategies of curriculum ideology and politics in the new era [J]. Journal of Xinjiang Normal University (Philosophy and Social Sciences Edition), 2020, (2): 50-58.
5. Yan Xianjun. On the "ideological and political elements", practical misunderstandings and educational evaluation of "curriculum ideological and political" in colleges and universities [J]. Ideological Education Research, 2020, (2): 88-92.

Research on a New Mode of Oral English Teaching Based on Computer Network Assistance

Yajie Liu

Zhengzhou University of Science and Technology, Henan, Zhengzhou 450064, China

Abstract. Oral English teaching is an important form of cultivating students' English accomplishment and improving their English learning effect. In the context of intelligent education, by giving full play to the advantages of computer network and actively reconstructing the new mode of Oral English teaching, we will, on the basis of optimizing the form of oral English teaching and by integrating high-quality oral English teaching resources, to promote students to participate in oral English learning and speaking practice in a conscious and spontaneous way. Based on the analysis of the existing problems in spoken English teaching, this paper puts forward the construction and implementation of a new spoken English teaching model based on computer network.
Keywords: Computer, Network-assisted, Oral English teaching, New model.

The convenience of network environment can greatly assist oral English learning. Language learning process is the input of information, through processing, thought transformation, and then output, so as to achieve communication. With the aid of computer network, students can input and output conveniently. In listening, speaking, reading and writing these English skills, the speaking part, that is, speaking, that is, the process of output. It's a difficult process. If you want to speak clearly and clearly, you need to practice with your heart in the process of accepting input and get more exposure to the English environment.

1 Construction background of computer network-assisted oral English teaching model

In the traditional English classroom, the classroom capacity is limited, teachers will not spend all the time in the classroom for oral practice, students' English learning also have to pay attention to listening, speaking, reading and writing aspects. Classroom resources are also limited, the use of classroom spoken language materials for oral practice for students, breadth and depth are far from enough, the urgent need to strengthen the edification of the English environment in daily life, strengthen oral practice. In the process of Oral English teaching, teachers also need to use a wealth of information to expand their oral practice. Teachers can use the internet to find information that is of interest to students. In class, teachers can ask questions about some topics related to students, create a harmonious and equal communication environment for students, let students open their mouths boldly, and use the network environment for teaching assistance, gradually improve the students' acceptance of oral practice. The network environment can also be used for efficient output. Under the guidance of Constructivist learning theory and cognitive theory, we explore a new oral English teaching model, which takes the network as the background, students as the center and communication as the focus. The teaching experiment is supported by computer network, and the teaching activities focus on information input, autonomous learning and communication training. Language teaching practice has proved that computer network-assisted college oral English teaching has achieved remarkable results. The experiment stimulated students' interest in learning English and improved their comprehensive ability of using English. The network environment helps the student to study at any time, as long as has the study idea to be possible to understand the related content on the network. In traditional oral English teaching without network environment, students have to study oral English in the language practice room or in the classroom, and the time and space are strictly limited in the classroom. With online learning, students can listen to English songs after class to develop their sense of language, and use English learning materials in their dormitory to develop their oral English habits, create English practice opportunities for yourself in your living environment.

Computer-based and web-based English teaching refers to the integration of various media and technologies into English teaching, and the storage, processing, and dissemination of English teaching and learning information, optimizing the information and teaching environment of English teaching and learning. On the basis of strictly following the modern teaching concept and making full use of advanced information technology, English web-based teaching attaches importance to creating an autonomous learning environment for students, emphasizing individualized teaching, forming an in-class and out-of-class, online and offline, self-study, face-to-face teaching and other interactive complementary student-centered teaching model. This change in the role of teachers and students, the new teaching mode, which emphasizes learning rather than teaching, is of far-reaching significance to the cultivation of students' abilities of comprehensive application of foreign languages and autonomous learning. However, in the process of English teaching, the leading role is still people, that is, the role of teachers can not be ignored. A violation of this principle is bound to have undesirable consequences. No matter how science and technology develop, no matter how society progresses, human language eventually needs to be taught.

According to the actual situation of English teaching and the characteristics of network teaching reform, the primary network teaching mode, which is classroom-based and computer-aided, has been popularized. This form of English teaching is still classroom-style teacher teaching, different from the traditional, there are more computer network AIDS. With the development of English teaching materials, teachers not only rely on sound, blackboard and other forms to impart knowledge, but also need to make use of computer digital technology to present these information more intuitively in the form of the network, it is convenient to carry out the teaching strategy of teaching students according to their aptitude, to give play to students' subjective initiative, to improve efficiency, and to fully realize the three-dimensional interaction between students and teachers.

In the design of test questions, authenticity is an important criterion of test questions design, that is, the test tasks are more and more close to the actual situation of language use. Traditional tests often break down language into sub-tests of knowledge and skills, while computer-based tests began to design comprehensive test tasks according to communicative tasks, speaking test tasks are as close as possible to the real English situations that students may use or can foresee in their future careers. Students should use listening, speaking, reading and writing skills to complete a specific task. To stimulate students' interest and motivation in learning English, the application of images, sounds, films and videos and other innovative test questions can stimulate students' positive emotions and motivation to communicate. In the traditional classroom, the teacher transfers knowledge to the student, the teacher occupies the leading position in the teaching, the student is in the passive position. The computer-based test breaks the limitation of "classroom + textbook", and adopts the teaching methods of "face-to-face teaching in class", "self-study under the guidance of teachers on the internet" and "interactive question-and-answer on the internet platform", gradually changed the traditional concept of learning.

2 Functional modules of computer network-assisted oral English teaching model

The network interactive platform of spoken English teaching is an open teaching supporting system. Whether its function design is reasonable or not will directly affect the teaching process and effect. A perfect computer network-assisted oral English teaching model should generally include the following four functional modules:

2.1 Teaching activity module

Teachers can edit videos of classroom activities into short videos and put them on an online teaching platform as teaching resources. Students can also upload their own multimedia videos onto the website, in this way, the network teaching platform has become a platform for students to learn and show themselves, will greatly mobilize the enthusiasm of students to learn and participate.

2.2 Video lesson modules

The module is mainly for students in the network teaching platform to watch video teaching content, according to their actual situation to choose their own learning resources. For those who have the ability to learn, they can learn high-level or extensive learning resources on their own, while students with learning difficulties can watch some content repeatedly.

2.3 Platform management module

The management module is the teacher and the platform manager to the teaching process and the platform function control. Therefore the network learning platform needs to provide the classroom management function module. When teaching, the teacher can control the status of other functional modules in this module. For example, when students are not required to discuss, the teacher can turn off the collaborative communication module so that students can not access the discussion page. This will enable students to focus, and teachers to keep pace with the teaching progress, improve the efficiency of online teaching.

2.4 Tool resource module

Oral English Teaching Network Interactive Platform is an open interactive website, can select network live television programs, through the campus broadband network connection related to the website online to watch the original film. Many web sites have rich, authentic language teaching materials and audio-visual aids, web-based teaching platform also has video search, popularity ranking, questionnaire and other site-specific tool module. Video search, mainly to provide a variety of ways to facilitate students to find all kinds of video resources; questionnaire survey, can be set regularly on the front page of some representative questionnaire, to improve the teaching of reference. In addition, there are a variety of digital, graphics, sound, images, animation and other multimedia tools, and can directly call MSN and QQ network video communication tools.

3 Strategies for constructing a computer network-assisted oral English teaching model

One of the biggest differences between this web-based teaching platform and other video English teaching websites is that students' performance in real classes is used as teaching materials and results, this is complemented by lectures by teachers and guidance from classic video materials. The participation of students in the whole teaching process is the key to the platform application.

3.1 Teachers and students work together to compile real-time classroom videos

The collection of real-time video in English class needs to track and record the whole class, and is targeted and thematic, therefore, the video collection staff on the collection of tools and the collection of video points for pre-training. And the video files collected in real time should be edited and optimized to make the classroom video better serve the students, teachers and students can be arranged to work together on the video file processing and processing. Voice, picture and body language can show the teaching content more vividly, strengthen the infection of teaching, and improve the teaching effect to a great extent. At the same time, the study also reminds that these teaching aids need teachers to give full play to their unique advantages, to stimulate students' subjective initiative, in order to really play an auxiliary role.

3.2 Integration of various modes of in-class and out-of-class activities

On the basis of listening and watching more online English audio-visual information, guide students to study independently or cooperatively, and exchange the video information orally with classmates. The teacher should arrange the students to practice oral expression in a planned way. The specific approach is: for in-class activities, there are several web-based activities of Oral English video teaching mode, namely: Story Reconstruction, film narration, simulation reporter interview and role-playing, scene Interpretation and debate competition, etc. . The teaching mode of extra-curricular activities includes film and TV program rehearsal and self-study. These activities can greatly stimulate students' interest, enable students to actively participate in teaching activities, and exchange discussions with teachers and students, from passive learning to active learning, so as to cultivate the students' oral expression ability. In addition, under the self-regulated learning model, students can use audio and video learning materials to learn on their own. After that, teachers can carry out in-class activities and carry out various interactive teaching activities, review and discuss student assignments. We should pay attention to the input of the cultural background of English-speaking countries, and let students have oral communication through living situations, materials and practices. With the development of language, students can understand the new trends of language development, place themselves in the culture behind the language, and gain more background information and deeper understanding of the language, enhance the real sense of the situation, stimulate the interest in learning, mobilize the desire for active learning, thus improving the effect of oral training. To construct a new model of oral English teaching, so that students can practice English in a good language environment, and finally achieve the goal of learning and using English.

3.3 Selection of video topics according to students' interests

For the assessment of Students' achievement, the assessment method is the combination of formative oral test and summative assessment. The aim of formative assessment is to encourage students to learn, to help students to control their learning process effectively, and to enable students to achieve a sense of achievement and self-confidence. In this way, students can not only measure what they have learned, but also strengthen and consolidate what they have learned, and apply what they have learned to real communication. Students also from passive acceptance of evaluation into the evaluation of the main and active participants. There are videos about campus life because they resonate with them, and there are videos about self-help. However, most of the students want to control the length of the clip in less than 10 minutes. According to repeated practice, the length of the clip should not be too long. The clip in less than five minutes is easier for teachers to explain and students to learn. Oral English teaching still has a wide range of research prospects, in the research methods and research focus is still waiting for researchers to open up new ideas, new perspectives. The deepening and perfection of the research on oral English assistant teaching by scholars will further promote the quality of oral English teaching in China, it will also bring benefits to the further development of English teaching and other language teaching in China.

4 Conclusion

Based on the image, lifelike and emotional situational learning and testing functions created by computer-assisted oral English teaching, students' language motivation and potential intelligence are stimulated, it can arouse the students' emotional experience, help them understand the test content quickly and correctly, promote the release and exertion of their comprehensive ability of language use, and improve the effect of Oral English learning.

References

1. Ye Rong. Feasibility and Approach to Infiltrate Humanistic Quality Education in Higher Vocational English Teaching [J]. Journal of Jiamusi Vocational College, 2015, (5). 10-11.
2. Xu Yan. An Analysis of Humanistic Quality Education in Comprehensive English Teaching in Higher Vocational Colleges [J]. Journal of Nanchang Institute of Education, 2013, (12). 119-120, 137.
3. He Min. An Analysis of the Cultivation Strategies of College Students' Oral English Communication Ability [J]. English Square (late issue), 2015, (4): 124-125.
4. Zhang Luan. An Analysis of the Cultivation Strategies of College Students' English Language Communication Ability [J]. English Square (Latest Issue), 2015, (9): 98-99
5. Li Jiaji. The Oral Competence Cultivation Strategy in English Teaching in Higher Vocational Colleges [J]. Campus English (Early Ten), 2015, (4): 73.

Construction Technology Safety Management under the Background of BIM and Information System Modeling

Yang Liu

Urban Vocational College of Sichuan, Sichuan, China, 610110

Abstract. The construction industry is the driving force and pillar of my country's economic development. In recent years, due to the increasing difficulty of construction and long construction period, the traditional construction technology safety management mode is no longer suitable for modern construction projects. Construction safety accidents also occur frequently and threaten The personal safety of construction workers has caused a serious negative impact on construction enterprises. The application of BIM technology and building information system has created a new management model for construction safety management in the construction industry, greatly improving the level of engineering construction safety management, and also marking the construction industry's march toward a new era of technical information.
Keywords: BIM technology, Information system, Safety accident, Construction technology safety management.

1 Introduction

Traditional engineering projects usually adopt a multi-person safety management model, but this will make building information distributed, making information nodes difficult to manage and screen. In the process of safety management, there will be a lot of information messy and undetectable hidden safety hazards. Problems such as construction technology and safety management can not be timely and accurately supervised and managed, which reduces the quality of construction and endangers the lives and safety of workers. The application of BIM information system in the process of building construction technology safety management reduces the occurrence of safety accidents and opens up new ways for construction safety management.

Many scholars at home and abroad have conducted research on the safety management of building construction technology under the background of BIM and information system modeling, and the research progress is relatively smooth. For example, a researcher studied the application of BIM technology in building information integration. It is very effective to integrate construction information with the help of BIM technology. After analyzing the informatization situation of the domestic construction industry, based on BIM technology, through a series of strategic guarantees such as talent development, main body coordination, and information exchange, a management model for targeted construction projects has been established [1]. A research team designed an innovative framework based on virtual reality technology to realize real-time data transmission from position sensors to visualization platforms, allowing users to be on the scene and dynamically monitoring the safety of the manufacturing process. They believe that the awareness of managers is in the construction The safety of the management structure in the process is very important. In the early stage of construction, safety construction designers need to make detailed safety plans to enhance the safety of the construction process [2]. Although the research results of construction technology safety management based on BIM and information system modeling are good, construction safety accidents still occur frequently, and effective measures need to be taken to reduce the accident rate.

This article introduces the advantages of the application of BIM information system in construction technology safety management. The implementation purpose of construction technology safety management based on BIM information system analyzes the types and locations of construction safety accidents in a province from 2015 to 2019, and uses BIM Technical management of the construction technology safety of the construction industry, and promote the healthy development of the construction industry.

2 BIM information system and construction technology safety management

2.1 Advantages of BIM information system

First of all, BIM information system has the advantage of reducing risks. As we all know, the construction process of a building is affected by different factors and usually faces different risks during the construction process. Some risks may hinder the timely completion of the construction project. In the traditional building design stage, designers must check and compare the construction design, but in the later construction process, there are still a large number of design modifications, which poses a major risk to the construction project. With the help of BIM technology, because the 3D model of the construction project can be presented more intuitively, the design accuracy of the construction project has been significantly improved, which not only reduces potential risks, but also allows designers to generate the best design ideas.

Secondly, BIM information system has the advantage of visualization. Visualization is one of the most basic capabilities of BIM information systems, and it is also the most obvious feature that differs from traditional architectural models. Essentially, BIM technology uses a three-dimensional method to present the characteristics of a three-dimensional building. It can not only enlarge and rotate, but also modify and enhance it. This is very different from the traditional two-dimensional design function. The two-dimensional method can only express the basic and key features, and cannot intuitively present the design information, so that the designer cannot find design errors and problems in time, which affects subsequent construction progress [3].

Furthermore, the BIM information system has the advantage of information exchange. Compared with traditional methods, one of the most important properties of BIM technology is the huge amount of information it contains, which also reflects the shortcomings of traditional methods that cannot analyze large amounts of information. Due to the limited methods of storing information, traditional methods cannot store large amounts of information. In this case, errors and problems when transferring information are very common. With the help of the BIM information system, various information can be retained from the initial design stage of a construction project, and it can be sorted and integrated with the information in the next stage of construction and operation to realize the penetration and exchange of information, while also ensuring the continuity of information.

Lay a solid management foundation for construction projects.

Finally, BIM information system has the advantage of parametric simulation. Parametric simulation is the cornerstone of a BIM information system and a prerequisite for realizing its series of advantages. In BIM technology, parametric modeling can be done through widgets and family libraries. This method can save the geometric and informal division parameter information of construction projects, which brings great convenience to designers [4-5] .

2.2 The basic process of building construction technology safety management

Building construction technology safety management is the process of designing, formulating and implementing safety decision-making plans based on the identification and measurement of factors affecting safety. Construction companies must follow the principle of "prevention first" and take targeted accident prevention measures for all safety issues involved in the construction process to ensure the personal and property safety of the engineering company and ensure the safe progress of construction. Therefore, the safety management task of the construction stage is mainly to predict the factors that affect construction safety, so as to take accident control measures to prevent the occurrence of safety accidents. Among them, the key is to establish a clear prediction and control system for factors affecting safety, and to integrate and systematically manage the safety of structures. Figure 1 shows the basic process of building construction technology safety management. First, determine the safety goal, that is, determine the safety goal of the construction project according to the actual construction situation (such as whether the construction site is a living house or a commercial house), and establish environmental safety protection and conduct safety education and training before implementing the construction; secondly, identify the impact on-site operation activities safety factors, find out all the factors that affect the implementation of the construction safety goals of the construction project; then, the project management personnel formulate and review risk mitigation measures and accident emergency plans to prevent and control major risks to ensure that construction safety issues are effectively resolved. Finally, they analyzed and evaluated the implementation results of the safety risk control measures during the construction of the construction project, and continuously improved the control measures to finally achieve the goal of construction safety. The key step that must be carried out is to classify and identify the factors that affect safety, which is a prerequisite to ensure the smooth implementation of construction safety work [6-7].

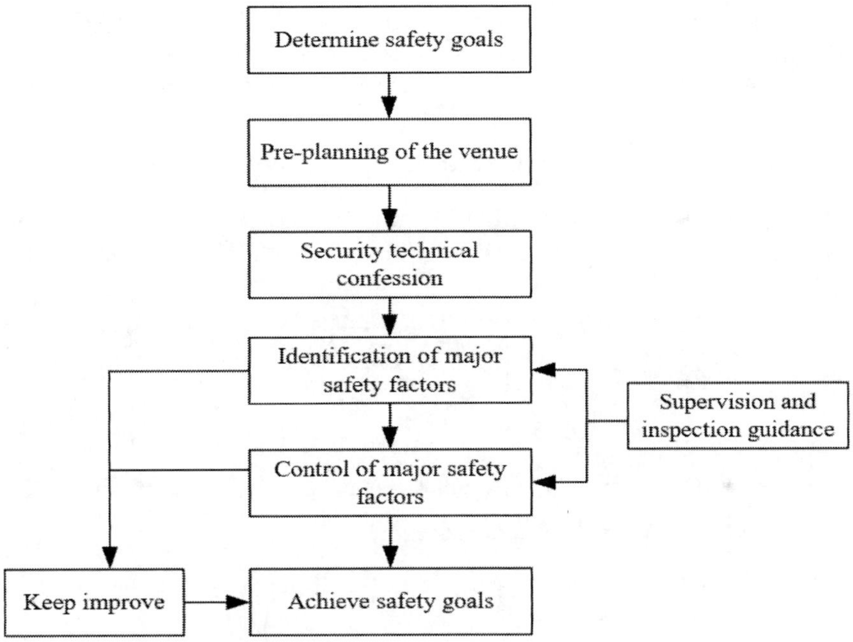

Fig.1. The basic process of building construction technology safety management.

2.3 Implementation purpose of construction technology safety management based on BIM information system

(1) Transparency of security status

The construction process is a dynamic process, which leads to uncertainty in the safety status of the construction site. It is the uninterrupted goal of all safety management personnel to maximize the safety of the construction site. Checking the information on the safety status of the construction site is closely related to the possibility of a safety accident. It is always related to the safety of the construction site and is also the focus of safety management. The implementation of on-site safety management based on BIM information system can help managers to understand the on-site safety status in real time, accurately and efficiently, so as to achieve the goal of improving the transparency of safety status [8].

(2) Intuitiveness of safety management

The construction technology safety management based on the BIM information system will improve the intuitiveness of safety management. Even beginners with little knowledge of safety can directly assess the construction site through the 3D display in the on-site assessment and the inspection and review of the transportation model. The safety status of the system allows them to have a comprehensive and intuitive understanding of the entire process, and also enables safety managers to effectively control the construction site [9].

(3) Dynamic safety management

In the process of construction safety management, BIM technology can monitor construction safety management rules in real time at any time of the day. In addition, the construction site is designed and organized in a 3D virtual scene, and the detailed construction plan is constantly updated and improved. Through dynamic simulation to ensure the continuity of on-site safety management in time and space, discover deficiencies and failures in time, and realize the dynamic construction of safety management.

(4) Proceduralization of safety management

The application of BIM technology in the construction process has promoted the implementation of safety management throughout the entire process from construction to completion. Add BIM technical data on the basis of the traditional management process, dynamically manage the entire construction process, standardize the management process, and complete the implementation of the construction safety management program [10-11].

2.4 BIM-based extraction of architectural topological relations

When extracting the spatial relationship within the building, each node in the topological graph is used as a starting point to calculate, and a series of spatial syntax variables are constructed [12]:

$$CON(u_i) = \sum_{j=1}^{n} e_{ij} x \tag{1}$$

Among them, e_{ij} is the edges of nodes u_i and u_j, $CON(u_i)$ represents the number of edges starting from node u_i, that is, the node connection value, and n is the number of nodes in the topological relationship graph.

$$I(u_i) = \frac{n-2}{2(MD(u_i)-1)} \tag{2}$$

Among them, $MD(u_i)$ is the average depth value of the node u_i, and $I(u_i)$ is the degree of integration.

3 Research on the safety management of building construction technology under the background of BIM and information system modeling

3.1 Research purpose

The BIM information system can use its visualization and simulation functions to detect certain safety issues in building construction, so as to prevent the emergence of safety accidents in the construction process in advance. This article attempts to explore how to use the building information system model to improve the level of construction safety management and effectively prevent and contain the occurrence of on-site safety accidents.

3.2 Research methods

This paper collects the frequency of accident types in construction safety in a province from 2015 to 2019, analyzes the location of safety accidents, and uses BIM technology to reduce the incidence of construction safety accidents.

4. Analysis of construction technology safety management

4.1 Statistics of construction safety accidents

Table 1. 2015-2019 Frequency of Construction Safety Accidents

	2015	2016	2017	2018	2019
Fall from height (%)	32.75	35.61	41.98	44.57	45.12
Collapse accident (%)	12.33	14.06	13.48	13.26	10.74
Object strike (%)	17.28	13.41	15.50	12.76	11.08
Lifting injury (%)	6.77	8.35	10.61	8.18	11.27
Suffocation by poisoning (%)	14.64	12.87	16.34	11.22	9.69

As can be seen from Table 1, there are five main types of construction safety accidents in the province from 2015 to 2019. Among them, falling from height is the risk factor with the highest frequency of safety accidents during construction. Falling from height in 2019 is safe the accident rate is as high as 45.12%; the rate of collapse accidents remains within the range of 10%-15%; the rate of safety accidents hit by objects is within the range of 10%-20%; the incidence of lifting injuries is the lowest, only 6.77% in 2015 ; the accident rate of poisoning and suffocation dropped below 10% in 2019, but it is still above 10% in other years. Therefore, corresponding preventive measures should be taken for these high-frequency construction safety accidents to minimize the number of accidents.

4.2 Locations where construction safety accidents occur

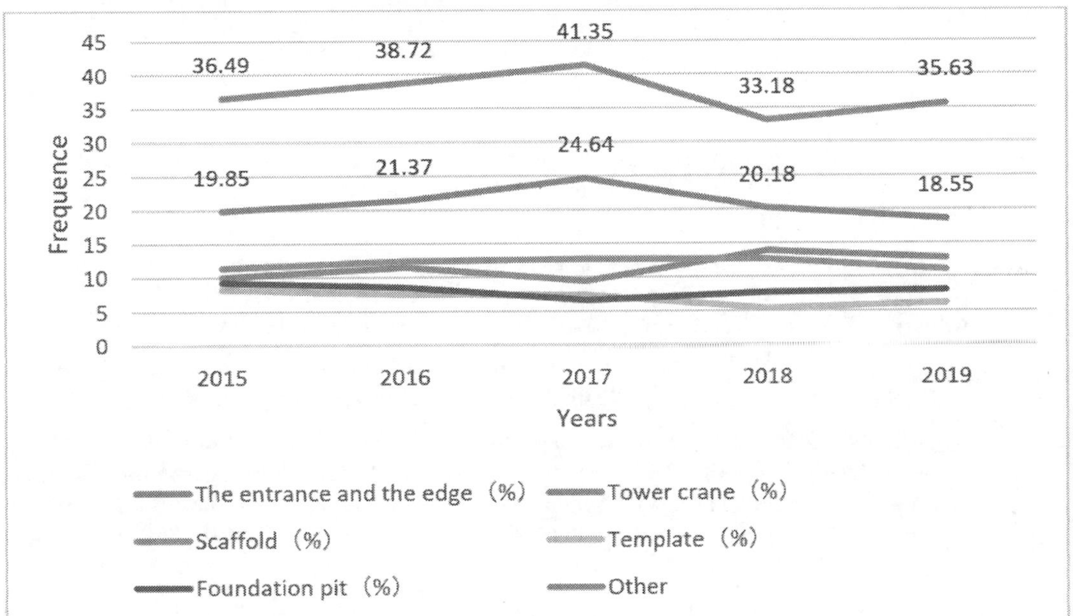

Fig. 2. Statistics of locations where construction safety accidents occur.

It can be seen from Figure 2 that the main parts of the province's construction safety accidents are: openings and borders, tower cranes, scaffolding, formwork, foundation pits and so on. The most obvious of which is the highest accident rate at the entrance of the cave and the edge, which has become the most difficult point in the safety management of construction. For these frequently occurring parts of safety accidents, appropriate technical safety management measures should be taken, such as reinforcing the construction parts of the scaffold to reduce the frequency of safety accidents due to the instability of the scaffold.

5 Conclusion

This article analyzes the incidence of construction safety accident types in a province in recent years, and understands where the safety accidents occur. In general, the reasons for the frequent occurrence of safety accidents during the construction process can be attributed to the following factors: First, people pursue the differentiated operation of the project, improve the building quality requirements and the existing building height, which greatly increases the difficulty of construction and complex buildings. The structure makes the occurrence rate of safety accidents higher; and many builders only pay attention to the progress and cost when managing the project, but do not pay attention to the management safety measures, and even reduce the safety management cost to complete the construction project ahead of time, without taking any safety protection measures , resulting in frequent safety accidents during construction and production; third, construction safety management methods are relatively backward, and many construction companies still adopt traditional management methods. Nowadays, with the passage of time, the scale of the project and the difficulty of construction have been greatly improved. It is no longer possible to realize the safety management of the construction site only by improving the management system and strengthening the management. In response to these problems, the use of BIM information systems to manage construction safety can effectively reduce the incidence of safety accidents.

References

1. Zhou Y. Research on Enterprise Green Management under the Background of Ecological Civilization Construction[J]. International Core Journal of Engineering, 2019, 5(11):130-134.
2. Deng L, Zhong M, Liao L, et al. Research on Safety Management Application of Dangerous Sources in Engineering Construction Based on BIM Technology[J]. Advances in Civil Engineering, 2019, 2019(3):1-10.
3. Li Peng An. Research on the Construction of Medical Information Sharing Platform under the Background of Large Data[J]. Modern Management, 2018, 08(1):68-76.
4. Olga Lucía Ostos, Rafael Rentería-Ramos, Cala F . Construction of a Complex Network Using Technological Surveillance for the Strategic Management of Science, Technology, And Research in Higher Education Institutions in Colombia[J]. Turkish Journal of Computer and Mathematics Education (TURCOMAT), 2021, 12(14):1185-1197.
5. F Kong, Li J, Lv Z. Construction of intelligent traffic information recommendation system based on long short-term memory[J]. Journal of Computational Science, 2018, 26(MAY):78-86.
6. P Mesáro, A Behúnová, T Mandiák, et al. Impact of enterprise information systems on selected key performance indicators in construction project management: An empirical study[J]. Wireless Networks, 2021, 27(3):1641-1648.
7. Love P, Zhou J, Matthews J. Project Controls for Electrical, Instrumentation and Control Systems: Enabling Role of Digital System Information Modelling[J]. Automation in Construction, 2019, 103(JUL.):202-212.
8. Love P E D, Zhou J, Matthews J, et al. System information modelling in practice: Analysis of tender documentation quality in a mining mega-project[J]. Automation in construction, 2017, 84(dec.):176-183.
9. Deng S, Zhang X, Shao Z, et al. An integrated urban stormwater model system

supporting the whole life cycle of sponge city construction programs in China[J]. Journal of water and climate change, 2019, 10(2):298-312.

10. Cho H P, Lim H, D Lee, et al. Patent analysis for forecasting promising technology in high-rise building construction[J]. Technological Forecasting & Social Change, 2017, 128(MAR.):144-153.
11. Ismail Z A. Optimising the safety of road transport workers on IBS building construction projects: a review[J]. Social Responsibility Journal, 2019, 15(6):837-851.
12. Chukwuma, Nnaji, John, et al. Improving construction work zone safety using technology:A systematic review of applicable technologies[J]. Journal of Traffic and Transportation Engineering(English Edition), 2020, v.7;No.37(01):66-80.

Analysis on the Integrated Accounting System of Enterprise Accounting and Statistics in the Big Data Era

Jun Liu

Zhejiang Open University Tongxiang College, Zhejiang, 314500

Abstract. In recent years, our country's comprehensive national strength has been continuously enhanced, the market economy system has been continuously improved, and the development of each industry has also undergone certain changes. The increasingly vigorous development of society and the increasingly fierce market competition are both opportunities and challenges for enterprises. If an enterprise wants to get better development, it is necessary to strengthen the management of its financial work. The current establishment of an integrated system of accounting and statistics is a good idea to improve the efficiency of corporate financial management and promote corporate development. Therefore, this article aims to study the integrated accounting system of enterprise accounting and statistics in the era of big data, and make full use of it on the basis of analyzing the status quo of our country's economic information system, the reasons for this status, and the integration of accounting and statistic accounting. The advantages of big data technology have built a new economic information production system based on big data, organically combined accounting and statistical accounting, and tested the functions of the system. The test results show that, the functional test finally found no bugs, and the performance meets the current needs of users and choices.

Keywords: Big Data, Accounting, Statistical Accounting, Integration.

1 Introduction

With the development of the market economy, accounting has developed from recording, measuring and reporting the results of economic activities to controlling and supervising the entire process of business activities, participating in business decision-making and long-term decision-making, and making contributions to macroeconomic management, and national regulations provide important information [1-2]. The characteristics and calculation characteristics of the statistical period and time point indicators with respect to numbers provide a theoretical basis for the difference between accounting stock accounting and flow accounting [3-4].

In recent years, in the operation and management of enterprises, they have gradually and rationally used the combination of accounting and statistics to perform financial accounting. This method can systematically, completely and uniformly grasp various quantitative relationships in economic operations. It is not only the core of the macroeconomic system, but also the basis for enterprises to achieve scientific management and decision-making [5-6].

Based on the analysis of the status quo of our country's economic information system, the reasons for this status, and the role of the integration of accounting and statistical accounting, this paper makes full use of the advantages of big data technology to build a new economic information production system based on big data. The accounting and statistical accounting are organically combined, and the function of the system is tested.

2. Analysis of the integrated accounting system of corporate accounting and statistics in the era of big data

2.1 The status quo of our country's economic information system

(1) The degree of information sharing is low. Information exchange between departments, departments and social stakeholders is even more difficult than during manual accounting. It not only wastes manpower and equipment, but also increases the amount of data, and the error rate reduces the timeliness of the use of accounting and statistical information, and its impact is much deeper than on the surface [7-8].

(2) The basic work of economic information is weak, causing information distortion and backwardness. Under the planned economy system, enterprises and grassroots organizations lacked strong demand for information, lacked the enthusiasm to collect and store their own and macroeconomic information, regarded the reporting of various reports as an additional burden, and lacked a sense of responsibility, benefit mechanisms, and legal obligations. Related to this, the manpower invested in information work is relatively weak and unstable, and the methods are relatively backward, making it difficult to accurately collect and transmit economic information in a timely manner, resulting in distortion and backwardness of macro-information [9-10].

(3) The content and methods of economic information collection and processing: The processing methods are backward and cannot meet the requirements of socialism and market economy. At present, the collected information is difficult to reflect the price information, value indicators, benefit indicators, supply and demand information of market economic activities, and there is a big gap; in terms of collection methods, the collection of major economic information such as statistics and accounting basically depends on comprehensive arrangements. The method of reporting level by level and summarizing level by level is cumbersome and cumbersome, with a long cycle and high cost, and it is difficult to adapt to the needs of the socialist market economy; in terms of processing methods, the overall level of modernization is low, and the grassroots is still mainly manual processing. Although statistics, accounting and other comprehensive departments are equipped with some computers, the software development, remote transmission, and database construction are not matched, and they lack uniform specifications and standards, resulting in low equipment utilization, low information development, and low sharing[11-12].

(4) In terms of the management system, the national economic information management agency is weak in organization and coordination. Government statistical agencies with relatively concentrated economic and social information are also unable to coordinate accounting, business accounting, and segmented departmental statistics due to institutional constraints, causing duplication, contradictions, closures, and loss of economic information, which affects the accuracy of economic information, sensitivity, systemicity, and relatively high cost.

2.2 Reasons for this status quo

(1) Traditional accounting and statistical theory

Theory comes from practice and serves practice as well as accounting theory. Accounting assumptions are based on a specific environment to better explain the nature, functions and basic requirements of accounting. When the objective environment changes, that is, when the assumed assumptions change, the accounting assumptions should also change accordingly. With the popularization of network technology and information technology, the emergence of virtual alliances and strategic alliances between companies has broken the scope of activities undertaken by the original accounting entities: online mergers and acquisitions, bankruptcies, etc. questioning the case of continuing operations. The emergence of financial statements can happen at any time, and the assumptions of the accounting period are useless; the emergence of electronic money and electronic banking makes people dream of paperless transactions and non-monetary measurement. This series of new changes will inevitably lead to changes in traditional accounting theories, and continue to be subject to traditional accounting theories, which will inevitably produce many problems.

Most of the country's statistical information comes from the report data of enterprises and the original vouchers of business activities. The transformation of traditional accounting theory and accounting information production mode also affects the accounting content and methods of statistics; in addition, many issues such as the globalization of the economy, the global nature of the network, and the blurring of industrial boundaries also affect the content, methods, and methods of statistical accounting and other theoretical systems. Today, when network technology and information technology are so developed, the traditional statistical theory system no longer meets the needs of modernization and is not compatible with more advanced productivity.

(2) In terms of individual information production mode

According to the traditional information production model, due to the limitation of technical means and the theoretical limitation of the "accounting object", when an economic activity occurs, the enterprise as a participant can perceive the occurrence of the transaction activity and only record information. The subject matter of the transaction, while ignoring the transaction link of economic activity. In this way, the company's accounting staff confirmed the company's internal financial operations and performed a series of tasks such as preparing coupons, writing account books, summary statements, tax returns, and external reports. From the above analysis, it can be seen that the traditional accounting information production model is an individual model, and most of the statistical data of the country's macro-audit comes from corporate statements and local summary statistics. After the company completes various statistical tasks, the local management department summarizes the statistical data, and then the management department summarizes the statistical data level by level, and finally forms the national macro-audit and financial management statistical information. In fact, the traditional statistical information production method is also an individualized and mechanically concentrated production method. The National Bureau of Statistics only conducts the final collection. Too many levels of centralization and reporting in the middle provide time and space for falsifying statistical information, which makes it difficult to coordinate accounting and statistics, and to share information, resulting in many difficult problems to solve.

2.3 The effect of implementing the integration of accounting and statistical accounting

(1) It can promote healthy and sustainable business growth. At present, the enterprise

management system is gradually being improved and perfected, and the business decision-making of the enterprise will be more scientific. For a long time, financial data summarized through accounting and statistical calculations have played a reference value for corporate management. For example, when performing accounting, it is necessary to measure certain contracts, purchases and other data according to the actual operating conditions of the company. It is necessary to allocate resources rationally and to ensure that the company's financial benefits are maximized.

(2) The combination of accounting and statistical accounting can continuously improve the quality of comprehensive information. The integration of the two can reduce the workload of financial personnel and statisticians, prevent repeated transmission of information, make the accounting cycle smaller than before, further improve the efficiency of information processing, and gradually improve the level of corporate governance. Accounting and statistical accounting work covers a large amount of information about the company's financial data. The integration of the two can make financial data more accurate and timely, thereby helping business managers make correct decisions and promoting business growth.

3 Experiment

3.1 Database management system

(1) The functional structure of the database management system

The main function of the database management module is to collect, organize, store, and transmit raw data (information), manage the data of the entire system, and realize information sharing and information automation and integrated production. Once all kinds of data are truthfully input from the network terminal, the database can be established and refreshed. This means that as long as the system defines the agreed data format, structure and corresponding processing procedures, the data will be automatically based on the mode required by the system. To carry out the assembly, instead of just organizing and managing the traditional conference data as the key element. Its functional structure is shown in Figure 1:

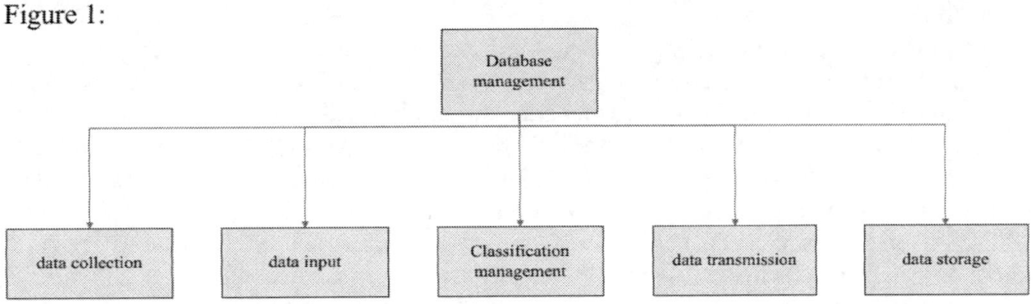

Fig. 1. Functional structure diagram of database management system.

(2) The main database

The databases used in SIPS mainly include the following categories: 1) enterprise-founded database; 2) accounting and financial database; 3) statistical information database; 4) industry contact database; 5) local transaction database; 6) remote transaction database; 7) Central transaction database; 8) GDP database; 9) Economic analysis database; 10) Input-output database; 11) Other economic data and economic analysis databases.

3.2 Economic entity establishment system

The main function of the establishment of an economic entity is to establish the establishment of an economic entity (including individuals), and allocate a certain amount of cyberspace to it, and grant certain permissions to it. A series of subsequent operations are based on its own cyberspace and data. And refresh the content in the database accordingly, and at the same time query and browse other information according to the permissions granted to the funds.

The establishment of economic entities generally includes the following contents: 1) Enterprises, including various enterprises engaged in economic activities; 2) Financial institutions, including banks, stock exchanges, etc.; 3) Government agencies at all levels, including governments at all levels from the bottom up Departments, such as the Taxation Bureau, the Statistics Bureau, the Industry and Commerce Bureau, and some other administrative agencies; 4) Intermediary agencies, including accounting firms, taxation firms, law firms, and other intermediary structures; 5) Individuals, anyone has to log in to the system, set up his own account and password, and obtain the information he needs according to his own needs.

3.3 Information system processing subsystem

(1) The functional structure of the information processing subsystem

This system is the core part of the integration of accounting and accounting. It is mainly used to process the original data of all business activities carried out by the enterprise from the establishment of economic entities to the economic activities, and perform a series of processing such as accounting and statistical accounting. All kinds of economic information needed by different departments are sent to economic information networks at all levels at the same time for all regions to collect and inquire.

(2) System description

This system mainly includes the following modules: 1) entity creation processing; 2) daily transaction accounting processing; 3) accounting information processing; 4) statistical data processing; 5) special declaration processing. The functions and roles of each module are as follows:

1) Entity creation processing. The main function of this module is that after the economic entity has successfully registered, it will establish an account and database associated with it, and perform corresponding accounting and statistical processing in its database, as follows: ①In the accounting database. The newly-built enterprise establishes accounting accounts and account books, and performs accounting initialization data processing (preparing vouchers, registering account books, financial reports); ②Establishing a statistical data processing account in the statistical information database for initial processing of statistical data. Technically, it can be carried out by establishing a sub-directory (folder); ③At the same time, record the relevant indicators of the economic entity in the regional economic information network, such as registration place, registered capital, etc., so as to facilitate related aggregation and statistical processing.

2) Daily transaction accounting processing. The economic transaction data in the to-be-processed database needs to be processed through the daily transaction accounting function to realize the processing of the accounting data (preparing vouchers, registering account books, financial reports), and generating the accounting data of each business entity. This is the data processing work of this system Important part.

3) Accounting information processing. On the basis of the above functions, the financial reports of each operating entity are compiled regularly for inquiries and use by information users who are authorized to inquire. For example, the financial reports of listed companies

are for public inquiry and can be sent and placed on the public information network for free inquiry. The financial reports of general enterprises are only open to authorized persons.

4) Statistical data processing. The part of enterprise statistical data derived from economic transaction data is stored in the to-be-processed database, which can be used for statistical processing, so as to realize the unification and information sharing of enterprises. Other information that needs to be specially declared by the business entity can be completed through the special declaration processing module.

5) Special declaration processing. In order to deal with issues like inventory consumption of production materials, or other information that requires special declarations by business entities, consider setting up this special declaration processing module.

3.4 Economic information analysis system

The role of the economic information analysis system is first to analyze economic information. Due to the advantages of the big data environment, accounting statistics can be integrated and socialized, economic information can be shared to a greater extent, and transparency and authenticity can be increased. So it can bring out the value of economic information to a greater extent. This system mainly uses the various basic data formed before, based on data warehouse, data mining, OLAP/OLTP and other advanced technical means, to analyze the accounting information and statistical information of enterprises, regions and even countries, and calculate various indicators such as GDP, etc. In order to dig out more valuable unified information, it is also conducive to the economic decision-making and macro-control of enterprises, regions and even countries.

3.5 K-means data mining algorithm

Cluster analysis is an important part of people's daily life, and it is also applied to various fields, such as business, education, Internet, and so on. Through clustering, people can discover which things are clustered together and which things are more similar, so that things can be well summarized and concentrated, and people can dig out some valuable and meaningful information from it.

Use formulas to describe the similarity of data in interval scale variables. The most commonly used distance formulas are Euclidean distance and Manhattan distance.

The Euclidean distance formula is shown in formula (1):

$$d(i,j)=\sqrt{(x_{i1}-x_{j1})^2+(x_{i2}-x_{j2})^2+\ldots+(x_{in}-x_{jn})^2} \tag{1}$$

The Manhattan distance formula is shown in formula (2):

$$d(i,j)=|x_{i1}-x_{j1}|+|x_{i2}-x_{j2}|+\ldots|x_{in}-x_{jn}| \tag{2}$$

4 Discussion

The whole test process of the system is iterated three times. In the iterative integrated function test phase, the total number of bugs tested is 28, 3, and 0 during the three iterations, respectively, showing a downward trend. The specific test summary of each module is shown in Table 1.

Table 1. Module bug statistics.

Test function module	Number of integrated function test bugs	Number of rectified bugs	Rectification ratio
Database management system	3	3	100%

Economic entity establishment system	4	4	100%
Information System Processing Subsystem	13	13	100%
Economic Information Analysis System	9	9	100%
Summary	29	29	

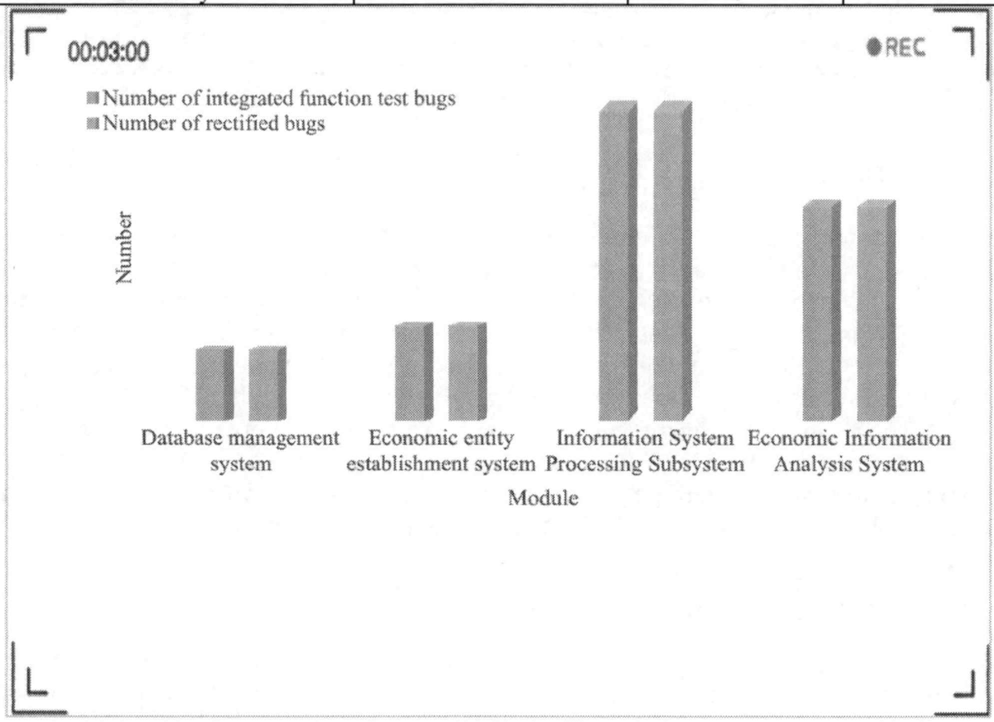

Fig.2. Module bug statistics.

As shown in Figure 2, in the testing phase, the system strictly follows the software testing process. The entire testing process is an iterative process. In the end, no bugs are found in the functional testing, and the performance meets the current user volume and selection requirements.

5 Conclusions

The continuous improvement of our country's financial system has put forward higher requirements for corporate financial management. If an enterprise wants to gain a firm foothold in the competition, it must start with financial management, innovate the financial management system, improve the integrated accounting system of accounting and statistics, and improve the quality of accounting personnel in order to maximize the financial benefits of the enterprise.

References

1. Bushman R M, Lerman A, Zhang X F. The Changing Landscape of Accrual Accounting[J]. Journal of Accounting Research, 2016, 54(1): págs. 47-78.
2. Agrawal A, Cooper T. SSRN-Corporate Governance Consequences of Accounting

Scandals: Evidence from Top Management, CFO and Auditor Turnover by Anup Agrawal, Tommy Cooper[J]. Quarterly Journal of Finance, 2017, 7(1):1650014.1-1650014.41.

3. Caballero R J, Farhi E, Gourinchas P O. Rents, Technical Change, and Risk Premia Accounting for Secular Trends in Interest Rates, Returns on Capital, Earning Yields, and Factor Shares[J]. Cepr Discussion Papers, 2017, 107(5):614-620.
4. Duan N, Liu X D, Dai J, et al. Evaluating the environmental impacts of an urban wetland park based on emergy accounting and life cycle assessment: A case study in Beijing[J]. Ecological Modelling, 2017, 222(2):351-359.
5. Li K K, Sloan R G. Has goodwill accounting gone bad?[J]. Review of Accounting Studies, 2017, 22(2):964-1003.
6. Kanodia C, Sapra H. A Real Effects Perspective to Accounting Measurement and Disclosure: Implications and Insights for Future Research[J]. Journal of Accounting Research, 2016, 54(2):623-676.
7. Shipman J E, Swanquist Q T, Whited R L. Propensity Score Matching in Accounting Research[J]. Accounting Review, 2016, 92(1):213-244.
8. Sachindra D A, Perera B. Statistical Downscaling of General Circulation Model Outputs to Precipitation Accounting for Non-Stationarities in Predictor-Predictand Relationships[J]. PLoS ONE, 2016, 11(12):e0168701-.
9. Chen F, Xue Y, Tan M T, et al. Response to comment on 'Efficient statistical tests to compare Youden index: accounting for contingency correlation'[J]. Statistics in Medicine, 2016, 35(4):637-640.
10. Henriksson M, Corino V, Srnmo L, et al. A Statistical Atrioventricular Node Model Accounting for Pathway Switching During Atrial Fibrillation[J]. IEEE Transactions on Biomedical Engineering, 2016, 63(9):1842-1849.
11. Henriksson M, Corino V, Srnmo L, et al. A Statistical Atrioventricular Node Model Accounting for Pathway Switching During Atrial Fibrillation[J]. IEEE Transactions on Biomedical Engineering, 2016, 63(9):1842-1849.
12. Hoang T V, Wu L, Golinval J C, et al. Stochastic multiscale model of MEMS stiction accounting for high order statistical moments of non-Gaussian contacting surfaces[J]. Journal of Microelectro mechanical Systems, 2018, 27(99):137-155.

Application of Multimedia Network Technology in Preschool Education Informatization

Yang Yu

Inner Mongolia Normal University, Inner Mongolia, Hohhot, 017000

Abstract. By means of information, fully integrating high-quality teaching resources and promoting the construction of pre-school education information, it is not only a change of promoting education and teaching mode, but also a vivid embodiment of the scientific upgrade of the concept of home interaction. In the informationization of pre-school education, the full application of multimedia network technology has broadened the space of pre-school education, built a home-based cooperative bridge, and created a new intelligent pre-chemical education environment, the ecology of preschool education has been completely reconstructed. This paper analyzes the application value of multimedia network technology in preschool education informationization, and puts forward some concrete application strategies.

Keywords: Multimedia; network technology; preschool education; Information; Application Strategy

In the informationization of preschool education, the full application of multimedia network technology will improve parents'information literacy on the basis of further strengthening the cooperation between kindergartens and parents, to help young children to develop good habits and positive attitude in the application of information technology, and promote preschool education to achieve high-quality, intelligent development.

1 The application of multimedia network technology in preschool education informationization

Multimedia technology can make children experience a more vivid audio-visual image in the learning process, teaching content more infectious, and further enhance the children's learning interest and learning enthusiasm. In order to let the child can understand the kindergarten clearly, the child care teacher may use the information multimedia technology to make and the teaching content related courseware. For example, play the kindergarten lawn, restaurants, playrooms and other places of video material. Then the teacher can show the video and explain it to the child at the same time. The child's attention is attracted by the content shown and the teacher's explanation. This teaching mode directly stimulates the child's audio-visual organ, children will be able to more directly understand the style and features of the kindergarten, the corresponding interest and learning enthusiasm will be promoted.

The application of multimedia technology in preschool education makes the difficult teaching contents easier to understand and the more abstract problems more concrete, the teaching difficulty and the key point more concrete, the visualization and the visualization displays, promotes the young child to carry on the observation and the understanding better. For example, in order to give young children a clearer experience of the hard work of workers and to cultivate their love of work, teachers first let young children discuss in small groups and ask them who built the beautiful lawn in the garden, teachers can then use multimedia technology to show children videos of workers cleaning garbage, building and mowing lawns, and then ask children to observe and discuss and answer the question. In this process, the teacher should interact with the child, after receiving the feedback from the child, the teacher should give full encouragement and affirmation, so that the child will form a good emotional experience.

Today's education requires that the principal position of young children should be highlighted and respected in the actual teaching process. Educators should further mobilize children's subjective initiative and stimulate their enthusiasm and enthusiasm for learning, let the child become the main body of the classroom, can study and explore independently. The application of multimedia technology in teaching can not only make the key and difficult points in teaching easier to solve, but also have a positive impact on the broadening of children's vision and the diffusion of children's thinking. At the same time, the use of multimedia technology, children can better participate in the classroom to play their best learning state. For example, when a teacher teaches a class about the beautiful homeland and perception of the hometown, the hometown emotion is difficult for young children to understand, and children's minds are still developing at the stage of preschool education, if the teaching content is abstract, then even if the teacher has the superb teaching art, also can not solve the teaching difficulty effectively. But through the use of multimedia technology, to provide children with rich image materials, and then supplemented by music and teachers' explanations, children will have an immersive emotional experience, and they will have a strong interest in their hometown, it is easier to learn this part of the content and the teaching effect is improved.

2 The application principle of multimedia network technology in preschool education informatization

In the use of multimedia technology should have a clear understanding, the use of its purpose and the process of integration with kindergarten teaching to think rationally. In order to give full play to the advantages of multimedia information technology, the following basic principles of applying multimedia technology in kindergarten teaching are put forward.

2.1 Scientific principles

The integration of multimedia technology and kindergarten teaching is not simply a make-do, but an organic combination of technology and teaching, which requires teachers to integrate multimedia technology organically into the teaching process and realize the functions of various teaching factors, in the teaching process to play a natural auxiliary role of multimedia technology, rather than rote use of multimedia technology, that is, technology for technology. On the one hand, teachers should select and use multimedia courseware scientifically. Teachers should choose the courseware according to the teaching content and the teaching goal, and apply these suitable courseware to the teaching activity through careful teaching design. In the application process, we should also pay attention to the adaptation and secondary processing, because a completely fixed courseware in the classroom is lack of vitality, teachers must make creative revisions and edits based on the teaching content and the actual situation of the children in the class. On the other hand, teachers should strengthen the consciousness of teaching design. In the teaching design of multi-media course, we must have the consciousness of integrating information technology with teaching. We can not use multi-media technology as a demonstration tool and simply use it to "create an ideal teaching environment", instead, we should adhere to the teaching concept of people-oriented, and carefully analyze the teaching contents, teaching objects, multimedia technology and other teaching elements involved in the teaching activities, pay attention to consider the function and characteristic of each element synthetically, avoid mechanical combination, repeated combination or blind combination.

2.2 The principle of purposiveness

The application of multimedia technology in the teaching process of kindergarten is to make multimedia technology serve the teaching, promote the reform of teaching and learning methods, and make it not only beneficial to the teachers' teaching, but also to the children's learning. Therefore, teachers should make it clear that multimedia technology can only be used as a teaching aid in teaching activities, and must overcome the dependence of the whole teaching activities on multimedia technology, avoid simply using new technology to move books and blackboards onto the screen, thereby simply turning "human" into "machine". Only by properly handling the relationship between multimedia technology and kindergarten teaching, making various teaching methods complement each other, and using multimedia technology in key points, can we finally achieve the goal of improving teaching effect. It is worth noting that the use of multimedia technology is also based on the needs of teaching, kindergarten teaching is a lot of teacher-student interaction, not every teaching activities need to use multimedia technology.

2.3 Developmental principles

As we all know, with the development of liberal education, the integration of multimedia technology and teaching has become a hot topic in the current curriculum reform. Different from the curriculum of primary and secondary schools, kindergarten curriculum has the characteristic of "life-like", which is the sum of all kinds of activities of children. At the same time, the biggest characteristic of modern information technology is its rapid development, teachers should realize that the integration of multimedia technology and kindergarten teaching is not a static process, but a dynamic process of continuous development and change, this requires teachers to firmly establish the concept of "child-oriented development", constantly update the concept of information technology, flexibly apply and innovate multimedia teaching methods, and make good use, flexible use and practical use of multimedia technology, give full play to the role of information technology in promoting kindergarten curriculum reform.

3 The application of multimedia network technology in preschool education informatization

It is an important goal of kindergarten education to cultivate children's interest in learning, their interest in feeling and expressing beauty, and their basic ability of using sense organs correctly. In today's rapid development of the information age, the application of multimedia technology has greatly promoted the realization of the above-mentioned teaching objectives.

3.1 Arouse interest and children's desire to learn

Psychological research shows that 3-6-year-old children tend to focus on unintentional attention, attention is unstable, accompanied by emotional progress, dull pictures, word-of-mouth teaching mode, it is difficult to attract children's attention. For young children, interest is often their direct motivation to learn. However, interest and motivation are not inherent, but through the novelty of external things, unique to meet the psychological needs of children to explore caused. The rational use of multimedia can better stimulate children's interest in learning, only when children have a strong interest in activities, the brain in a state of excitement, children will actively participate in activities. For example, the teaching purpose of the children's song "Clouds and wind" is to stimulate children's interest in beautiful literary works and cultivate children's ability to appreciate them, develop young children's imagination and language skills. The dynamic effect is more in line with children's hyperactive and curious psychology, which naturally stimulates the interest of learning, thus further guiding the children to the next step of learning, to achieve the goal of teaching. Most children like to watch TV animation is actually the psychological interest and curiosity to attract the child's attention. It is a great advantage of multimedia technology to arouse children's learning interest, enthusiasm and initiative, and make children study happily in a lively atmosphere.

3.2 Create situations to enrich children's thoughts and feelings

The use of multimedia technology can create a relaxed and pleasant learning atmosphere for young children and create a lively experience situation, which is conducive to stimulating their interest in literary and artistic works and natural science, so that children feel the works of art and the love of nature, awe, from which received emotional education. In this regard, there have been many kindergartens have tried, practice shows that this is an effective way of education. Taking ancient poetry education as an example, many ancient poems praise the unique scenery and scenery of the season. In order to enable young children to deepen their impression of the scenery and the understanding of the poet's feelings, making use of multi-media to add pictures to these ancient poems to make them become a vivid video, creating various poetic and picturesque situations contained in the ancient poems, and reducing the poet's emotions into a beautiful and vivid short story, improve children's understanding of ancient poetry.

3.3 Visual demonstration to promote children's understanding of teaching contents

Children's psychological and physiological characteristics determine that pre-school teaching activities should be adopted intuitive teaching methods. The introduction of multimedia into the classroom can realize the combination of words, images, sounds and animations, and make the original abstract and boring knowledge become vivid and vivid, especially in the aspect of cognition of the micro-world, it can make a huge difference. Take the teaching of common sense activities as an example, some natural phenomena are not easy to see at ordinary times, can not make children experience and understand, the key points and difficulties in the curriculum are not easy to break through. For example, the teaching goal of "the formation and harm of thunder and lightning" is to let children know that Thunder is a natural phenomenon, to improve children's awareness of preventing thunder and lightning injuries, and to master the basic knowledge of preventing thunder and lightning injuries. This lesson teaching relies on the traditional teaching pattern to achieve the teaching goal very hard. First, through organizing and watching the video materials such as "the birth of thunder and lightning", "taking shelter from rain under big trees" and "treading water in the fields", the teachers set up a scientific point of view by giving young children a preliminary understanding that thunder is a natural phenomenon, at the same time, I also know the danger of thunder and lightning and set up a sense of self-protection.

3.4 Lively and lively to promote children's sensory development

Maria Montessori has developed a number of sensory education textbooks and teaching methods to help young children develop sensory ability, that is, to provide brain stimulation to develop intelligence. Most children like to watch animation, animation is actually the video, music or language on the children have a stimulus, and easy to follow the program to learn the contents. Take the common sense activity "know the rainbow" as an example: it often appears in the summer after a thunderstorm, the clouds in a large number of water gathering, through the refraction of sunlight, there will be a beautiful rainbow in the sky. According to their identification, playing the rainbow component courseware made by the teacher in advance, the red, orange, yellow, green, blue, blue and purple colors can be shown in turn, and finally the seven-color rainbow can be formed, at the same time guide students to pay attention to the order of the various colors. The repeatability of multimedia allows children to watch the process over and over. Watch the end of the children can be organized to draw their own rainbow with a color pen, so as to achieve the effect of exercise ability.

4 Conclusion

Preschool education is a special education cause, kindergarten is the starting point of every child's education, all kindergarten teachers play a role as a life teacher, which is very important for children's learning and growth. Therefore, as a teacher, one must have a relatively high comprehensive quality and be able to give full play to oneself professionally. At the same time, a kindergarten teacher must keep pace with the times and be constantly integrated with the times, the education informationization and the preschool education effective union, realizes the preschool education informationization, realizes the teaching to face the future.

References

1. Xing Xishen, Xu Lin. Research on the development path of preschool education informatization in the 2.0 era [J]. China Electronic Education, 2019, (5): 49-55.
2. Qu Tiehua, Wang Hongjing. Changes in my country's preschool education system: trajectory, logic and trend [J]. Education Science, 2020, (3): 85-90.
3. Xing Xishen, Jin Chuanyang. Research on the modernization of preschool education boosted by informatization [J]. Modern Educational Technology, 2020, (6): 108-113.
4. Ni Beijun. Construction of Preschool Education Practice Teaching System from the Perspective of "Internet +" - Comment on "Preschool Education Practice Teaching Organization and Guidance" [J]. Research in Science and Technology Management, 2021, (5): 230.
5. Fu Weidong, Zhou Wei. The construction of preschool education teachers in my country during the "14th Five-Year Plan" period: the main situation and key tasks [J]. Modern Education Management, 2021, (4): 83-91.

Mental Health, Cognitive Ability and Family Income—Empirical Analysis Based on the Follow-up Survey of Chinese Households

Yizhou Gu

Department of Statistics, University of Warwick, Coventry CV4 8UW UK

Abstract. This article uses the 2018 Chinese Family Tracking Survey data to empirically analyze the impact of mental health on family income and explore the mediating role of cognitive ability. The results show that for every 1% reduction in the family's mental unhealthy degree measured by the CES-D8 score, the family income level will increase by 1.7%-1.9%. After adding the pre-determined explanatory variable 2016 mental state level and other operations, use the filtered loneliness As an instrumental variable, the same conclusion was obtained by performing 2-stage least squares; the results of indirect calculation show that the mediating effect of cognitive ability is about 16.96% of the total effect of mental health affecting family income level; in terms of heterogeneity analysis, Researched that the impact of mental health problems at different family income levels is different. The marginal effect of the mental health impact of low-income families is as high as -2.9%. The mental health status of the family finance chief has a significant impact on the income of the entire family , Especially low-income families.
Keywords: Mental health, Family income, Cognitive ability, Quantile regression.

1 Introduction

As an important part of human capital, health has an important impact on the performance of the labor market [1]. In recent years, the social economy has developed rapidly, the level of medical technology has improved significantly, and people's physical health has been rapidly improved. In 2016, the healthy life expectancy of newborns in the world has increased to 63.3 years, and the healthy life expectancy of newborns in China has increased to 68.7 years [2]. However, compared to the optimistic physical and healthy development, people are facing more and more prominent mental health problems caused by pressure in the fast-paced modern life. It is estimated that one-third of the world's people will have a mental disorder during their lifetime, and one-third of the patients treated in general hospitals are patients with mental disorders [3]. In this context, the mental health of family leaders has aroused widespread concern.

Especially in developing countries like China, mental health problems are particularly prominent. The "Chinese National Mental Health Report (2019-2020)" issued by the Institute of Psychology, Chinese Academy of Sciences and other departments on March 1, 2020 shows that [4] that mental health problems of different populations in my country are on the increase, and most residents are irritable, depressed, and depressed. Inferiority complex, loneliness and other negative emotions are troubled, and 11%~15% of the people have poor mental health. Psychological research shows that the physical and mental damage caused by depression, anxiety, and emotional instability can shorten the working hours of patients, reduce work ability and work efficiency, and then participate in labor with psychological problems, and reduce the patient's Labor income [5]. In addition, mental health is not only detrimental to the accumulation of family wealth and reduces labor supply in the labor market, but also causes huge family costs such as treatment costs and indirect costs. Mental health issues are not conducive to the optimization of personal welfare. It will also hinder the growth of corporate efficiency, increase social costs, and harm the welfare of the entire society. This shows that it is very important to pay attention to the economic impact of mental health on individuals, families and society.

In a research report "Mind the Income Gap", it was pointed out that mental health brings a huge income gap. People with mental health diseases such as anxiety or depression have an average annual income of 8,400 pounds less than healthy people[6]. There are three systemic reasons: low employment prospects, low wages, and long-term benefit cuts. At the same time, previous studies have shown that when someone in the family has mental health problems, it often brings negative emotions to the whole family, which is the process of emotional transmission [7-8]. This kind of emotional transmission will affect the emotions of the person in charge of family finances to the income level of the entire family. Therefore, in the context of the increasing influence of mental health on the quality of life of residents, a systematic investigation of how mental health affects family income and analysis of its internal mechanisms have important theoretical and practical significance, as well as improving residents' mental health and people's lives. It provides a useful reference for the formulation of relevant policies to reduce the gap between the rich and the poor. The structure of the subsequent parts of this article is as follows: The second part introduces the research progress of existing related literature, the third part introduces the empirical strategies and descriptive statistical results used in this study, and the fourth part introduces the main empirical results of this paper, including basic model regression Results, instrumental variable regression results, quantile regression results, the fifth part is the research conclusion.

2 Literature review and research hypothesis

Bogan and Fertig used American Health and Retirement Survey data to empirically analyze the impact of American families' mental health status on investment choices [9]. The results show that mental health, like physical health, has an important impact on investment decisions, and that households suffering from various mental health problems are likely to hold assets down by 19%. This confirms the theory that mental health issues affect family financial income by changing the individual's optimization ability. Thomas found in a longitudinal study of economic difficulties and mental health in a national sample of British college students that anxiety and alcohol dependence lead to a deterioration of economic conditions, and economic difficulties will lead to poor mental health of students, creating a vicious circle [10]. According to a new study, low family income levels are associated with several lifetime mental disorders and suicide attempts, and reduced income is associated with increased risks of anxiety, drug abuse, and mood disorders. It can be seen from the above that although there is a two-way causal relationship between mental health and income, from the perspective of government policy formulation, people's mental health cannot be improved directly by increasing income. From the perspective of health economics, this is a healthy choice process. During this process, individuals with mental health problems are less likely to complete education or engage in higher-income-related occupations. Therefore, intervention measures should focus on improving mental health, breaking the vicious circle, bringing confidence to low-income people, and enabling individuals to obtain higher incomes.

In the link where mental health affects family income, cognitive ability often plays an extremely important role. The relevant theoretical and empirical research on cognitive ability on income has been relatively rich. In the LMIC study, the cognitive score increases by one standard deviation, salary Would increase by 4.5%. Ozawa and others also pointed out that improving cognition is the key to improving economic returns, especially in low-income countries [11]. Lin et al. pointed out that a high level of cognitive ability can also increase the expected average lifetime income [12]. At the same time, cognitive ability not only affects wage income, but also affects other sources of income such as property income [13]. And mental health problems will bring huge obstacles to the cognitive process, and the impact on people's perception, attention, learning, memory, reasoning and problem solving is not only short-term [14]. Therefore, cognitive ability will be a very important mediating variable. This article will follow the logic chain of "mental health-cognitive ability-family income" to explore the mediating role of cognitive ability in mental health affecting family income. Low- and middle-income families may be less concerned about mental health problems, and their reluctance to consult and treat makes the impact of mental health on family income more obvious. Therefore, low- and middle-income families should be more concerned.

Based on the above discussion, the following hypotheses to be tested can be obtained:

Hypothesis 1: When other conditions remain unchanged, families with better mental health can increase family income;

Hypothesis 2: Cognitive ability has a mediating effect in the process of mental health affecting family income;

Hypothesis 3: The impact of mental health on family income is heterogeneous, and the impact of mental health of low-income groups may be more obvious.

3 Data, variables and measurement models

3.1 Data source and variable setting

The data used in this article comes from the China Family Panel Studies (CFPS) conducted nationwide in 2018. Since 2010, the permanent tracked objects have been tracked every two years, reflecting the community, family, Three levels of individual data. The data provides high-quality micro-data for studying the financial problems of Chinese households, including housing, financial wealth, consumption, income, etc. In 2018, the survey covered 14,217 households and 37,354 residents in 31 provinces (cities, autonomous regions) across the country. The CFPS data covers the information of each family member in the interviewed family. The "number of family members" ranges from 1 to 17. The remaining valid sample households are 456, including 3150 urban households and 1302 rural households. This article believes that the 2018 data is complete and the questionnaire structure is perfect, which is very representative for the study of household financial issues. To ensure the robustness of the results, this paper also includes 2016 survey data in the model for robustness testing.

3.1.1 Household income (explained variable)

The measurement index in this article is obtained from the adult database, which is the "family annual income", and the natural logarithm of the income is processed and added to the regression model.

3.1.2 Mental health (core explanatory variable)

The World Health Organization pointed out in 2001 that mental health is a state of health or happiness, in which individuals can realize themselves, cope with life pressure, work productively and productively, and can contribute to the society in which they live[15], Poor mental health is usually accompanied by a series of psychiatric problems such as depression, anxiety, fear, and excessive alcoholism [9]. This article uses a simplified CES-D scale to measure the mental health of the family, Turvey confirmed that this scale has good reliability and validity [16]. In the survey, the person in charge of family finance was asked how frequently the following behaviors or feelings occurred in the past week: (1) I feel depressed; (2) I feel strenuous to do anything; (3) I feel sad; (4) I feel lonely; (5) I feel happy; (6) I am happy in life; (7) I do not sleep well; (8) I feel that life cannot continue. The person in charge of family finance needs to choose among four options: (1) almost none (less than one day); (2) sometimes (1-2 days); (3) often (3-4 days); (4) Most of the time (5-7 days). According to the answer of the person in charge of family finance, assign 1 to 4 points to each question. After adding up, the initial value of CES-D8 score between 8 and 32 is obtained. The higher the score, the worse the family mental health.

3.1.3 Cognitive ability (mediation variable)

Cognitive ability. This article is based on the word analysis, mathematical calculation and memory ability scores of the interviewed family financial officers in the CFPS2018 questionnaire, as the three dimensions to measure the cognitive ability of the family financial officers, and sums the scores of the above three dimensions as An index to measure the individual's comprehensive cognitive ability [13, 17]. At the same time, in order to unify the unit of measurement and further clarify the size of the effect of cognitive ability on household income, this paper standardized the above scores in the empirical process [18].

3.1.4 Control variables

When explaining the variable of household income, this article mainly controls the personal characteristics and family characteristics of the person in charge of household finance. Variables related to the characteristics of the person in charge include: whether the respondent is male, 1 for male and 0 for female; age and the square term of age; education level: no education (0 points)-PhD (9 points); household registration Hukou type, agricultural hukou is 1, and non-agricultural hukou is 0; variables related to family characteristics include: family size, which is characterized by family population size; family property value is logarithmic; social willingness, measured by the interviewee's social time, If the respondent's social time is higher than the average, it is considered that the respondent has a stronger willingness to socialize. On the contrary, the respondent's social willingness is considered to be relatively weak. The corresponding score is 0-10; trust in society , Reflected from the side of trust in strangers, the degree is 0-10; this article also controls the province where the family is located. Table 1 reports in more detail the meaning and descriptive statistical results of the variables used in the empirical analysis below. Observing families who answered "don't know", "unsure", and null values were excluded.

Table 1. Descriptive statistics of variables.

Variable	Variable meaning	Mean	Standard deviation	Minimum	Max
Explained variable					
family_income	Family annual income (yuan) take the logarithm	10.793	1.160	0	16.03
Core explanatory variables					
mental_heth18	Family mental health, measured by CES-D8	13.8	4.228	8	32
mental_heth16		11.35	4.149	8	29
Mediating variable					
cog	Cognitive ability (normalized value of the sum of word ability, math and memory ability scores)	0.009	2.334352	-5.038162	6.435967
cog_word	Word ability	0.009	0.991	-1.872	1.52
cog_math	Mathematical ability	0.01	0.988	-1.661	3.386
cog_memory	Memory capacity	-0.01	0.996	-1.505	1.525
Control variable					
gender	Gender (male=1, female=0)	0.515	0.5	0	1
age	Age	53.486	13.472	18	93
age2	Age squared	3042.175	1447.341	324	8649
edu	The highest level of education, from "no school" to "doctorate", assigned values of 0-9	2.587	1.302	1	8
family_size	Family size (person)	3.759	1.887	1	17
registered_res	Country = 1, town = 0	0.727	0.445	0	1
heth	Health self-assessment: "Very healthy" to "unhealthy" are assigned a value of 1-5 respectively	3.227	1.218	1	5
house_value	Logarithm of property value	12.226	1.557	4.605	17.733
Inter_ac_inf	Internet access to information level 1-5 points	2.551	1.607	1	5
social_contact	Popularity score 0-10 points	7.153	1.974	0	10
social_trust	Score 0-10 points for trust in strangers	2.079	2.216	0	10

3.2 Measurement model setting

3.2.1 Regression model

$$Y_i = \beta_0 + \beta_1 mental\_heth18 + \lambda X_i + \varepsilon_i \quad (1)$$

Equation (1) represents the regression model used in this article,in Y_i Indicates the total household income of the respondent,$mental_health$2018Indicates the mental health of the interviewee in 18 years, who is also the person in charge of family finance,Its coefficient β_1It is the regression result that this article focuses on, representing the impact of mental health on family income; X_iIndicates a series of control variables used in this article, including: gender, age, age square, education level, family size, household registration, physical health status, logarithm of real estate value, degree of Internet access to information, social willingness, and social trust. Used to exclude the influence of other factors on the household income of the interviewee;ε_iIndicates random interference.

$$Y_i = \beta_0 + \beta_1 mental\_heth18 + \lambda X_i + \delta_i + \varepsilon_i \quad (2)$$

In order to further control the possible impact of unobservable regional factors on the regression results, this paper adds the province's location on the basis of formula (1) virtual variable δ_i get(2), It is used to control the difference in income levels of different provinces, so as to ensure the reliability of the estimation results.

3.2.2 Quantile regression model

Quantile Regression(QR) It was first proposed by Koenker and Bassett in 1978. This method can accurately explain the influence of variables on the variable range of the explained variable and the shape of the conditional distribution. It can fully describe all the conditions of the conditional distribution of the explained variable. It can also analyze the mechanism of the explanatory variable on the explained variable under the conditions of each quantile. Because the regression coefficient estimator under different quantiles is often different, it can explain the respective influence of the dependent variable on the independent variable at different levels. The quantile regression model used in this paper is as follows:

$$y_q(x_i) = x_i' \beta_q \quad (3)$$

in, β_qCalled q Quantile regression coefficient, Its estimator $\hat{\beta}_q$It can be defined by the following minimization problem:

$$\min_{\beta_q} \sum_{i:y_i \geq x_i'\beta_q}^{n} q\left|y_i - x_i'\beta_q\right| + \sum_{i:y_i < x_i'\beta_q}^{n} (1-q)\left|y_i - x_i'\beta_q\right| \quad (4)$$

Hypothesis q=1/2, It is the median regression. At this time, the objective function is simplified to:

$$\min_{\beta_q} \sum_{i=1}^{n} \left|y_i - x_i'\beta_q\right| \quad (5)$$

The median regression has the smallest absolute deviation estimator, making it less susceptible to extreme values than the mean regression, and the statistical results are more robust. In the following article, we will examine the regression characteristics of household income as 1/4 quantile, median, and 3/4 quantile, and compare them.

4 Analysis and discussion of empirical results

4.1 The impact of mental health on family income

This article first uses the OLS model to empirically test the impact of mental health on family income, and gradually adds control variables and regional fixed effects, and obtains the regression results shown in Table 2. Column (1) shows the univariate regression result that only considers mental health, that is, CES-D8 score. Considering the two-way causality between mental health and family income, the mental health status of the same family in 2016 is selected as the predetermined explanatory variable for the result In contrast, the two columns (3) and (4) show the regression results after adding the control variables and regional fixed effects.

Table 2. Regression results of the impact of mental health on family income.

	(1) family_income	(2) family_income	(3) family_income	(4) family_income
mental_heth18	-0.055***		-0.019***	
	(-19.74)		(-7.22)	
mental_heth16		-0.055***		-0.017***
		(-19.24)		(-6.42)
gender			-0.002	0.003
			(-0.11)	(0.14)
age			0.039***	0.040***
			(7.84)	(7.96)
age^2			-0.000***	-0.000***
			(-9.14)	(-9.24)
edu			0.123***	0.124***
			(13.66)	(13.59)
family_size			0.160***	0.160***
			(28.29)	(28.32)
registered_res			-0.392***	-0.391***
			(-15.70)	(-15.66)
heth			-0.019**	-0.023***
			(-2.13)	(-2.72)
house_value			0.217***	0.218***
			(27.06)	(27.12)
Inter_ac_inf			0.039***	0.039***
			(5.09)	(5.09)
social_contact			0.014***	0.017***
			(2.83)	(3.28)
Province difference	Control	Control	Control	Control
Number of observations	9291	9291	9291	9291
R^2	0.040	0.038	0.419	0.419

Note: The values in parentheses are standard errors,*、**、***They are expressed at the significance level of 10%, 5%, and 1%, respectively.

It can be seen from the results in Table 2 that the degree of mental health has a significant positive impact on family income. The higher the score of mental health, the lower the family income. Specifically, columns (3) and (4) of Table 2 show that after controlling for control variables such as personal factors, family factors, and regional fixed effects, the marginal effects of mental health impact on income are -1.9% and -1.7, respectively. %, based on the family mental health score measured by the CES-D8 score of 2018, each additional point will reduce their family income by 2.1%, while the mental health of the pre-determined explanatory variable in 2016 will increase by one point. Revenue will decrease by 1.9%. The difference may be due to the fact that the family's low income will affect the living conditions and quality of life of family members to a certain extent, increase family anxiety and dissatisfaction with work and life, and cause aggravation of mental health problems, that is, depression tendency and There is a two-way causal relationship between income. In order to solve this problem, we will construct instrumental variables in the follow-up to eliminate as much as possible the reverse causal interference of income on depression tendency. At the same time, these conclusions also preliminarily verified Hypothesis 1. In terms of control variables, gender has no significant impact on household income, which is different from some existing studies. Zhang Chuanchuan believes that when evaluating health intervention policies, it is necessary to comprehensively consider the impact of health changes on the social and economic status of citizens, and to consider men and women differently[19]; Wang Xiuzhi found that the impact of men's health on income is significantly higher than that of women, regardless of the reality In terms of physical labor or mental labor, men generally have an advantage over women [20]. Age and its square term have a significant impact on family income, which means that there is a more complicated non-linear relationship between the age of the person in charge of finance and family income, and the influence of age on family income has an "inverted U-shaped" effect [21].

Physical health has a significant impact on family income. The improvement of the health of financial managers will first increase healthy human capital and bring a positive income effect. At the same time, compared to "unhealthy" families, "very healthy" families have fewer living costs, such as medical care, insurance, and time costs. The positive impact of education and family size on family income is obvious. In order to try to control the characteristics of the person in charge of family finance, the positive feedback brought by the Internet and interpersonal communication is also consistent with the existing research.

4.2 Endogenous problems

Since there is usually a more obvious two-way causality between mental health and family income, there are potential endogenous problems in identifying the cause and effect of mental health on family income [22]. In order to eliminate possible predetermined endogenous explanatory variables and two-way causality problems, most current researches use instrumental variable methods to obtain more accurate estimation results. The aforementioned literature review summarizes the instrumental variables in the existing literature, pointing out the experience of suffering from mental illness in childhood, the mental health status of the three months before the survey, the previous psychological symptoms, the history of mental illness of the parents, and the past Whether a close friend has passed away in a year, religious beliefs, social support, physical exercise frequency, and stressful events can all be used as instrumental variables to measure personal psychological conditions. According to the availability of data, this article will screen and replace the variables in the relevant literature.

When performing instrumental variable regression, first perform the regression for the first stage, and select the exogenous variables that have the greatest impact on the endogenous explanatory variables as the instrumental variables to ensure that the instrumental variables used in this article have the best possible accuracy. Specifically, the instrumental variables and related definitions used in this article are shown in Table 3.

Table 3. Instrumental variables and related definitions.

Literature summary and possible instrumental variables	Definition of this article	variable name
Physical exercise frequency[23]	Past weekly exercise frequency	exercise
Religious belief[24]	Whether to participate in religious activities	belief
Resident mobility	Whether the place of residence at the age of 3 or 12 is the same as the place of birth	migration
Childhood experiences[24]	At the age of 3, at least one of the parents is absent from company (the absence of company for more than one month is defined as lack of company)	nofm3
	At the age of 4-12, at least one of the parents is absent from company (the absence of company for more than one month is defined as lack of company)	nofm12
Society and Family	1.Not hated by others	d_sick
	2.Close relationship with spouse	d_close
	3.Not alone	d_alone
	4.A sense of accomplishment	d_achieve

Among them, the influence of d_alone on mental health is better than other instrumental variables, as shown in Table 3. In addition, the F statistic of its one-stage regression is much higher than the critical value (about 16) recommended by Stock and Yogo to reject the hypothesis of weak instrumental variables at the 10% level [25]. Therefore, it can be considered that the instrumental variables used in this article have a strong explanatory power for the mental health of financial managers, and there is no problem of weak instrumental variables. It can be seen that the regression results of the second stage show that both the pre-determined mental state and the current mental state after the use of instrumental variables have a significant impact, that is, a high score (the worse the mental health) reduces the family income.

Comparing the coefficients of mental health in Table 2 and Table 4, it can be found that the degree of mental health has a greater impact on family income in the regression results after using instrumental variables. This may be because family income itself has a certain influence on the mental health of respondents. Therefore, the original regression results underestimated the positive impact of mental health on family income.

Table 4. Two-stage instrumental variable regression results.

variable	(1)	(2)	(3)	(4)
	mental_heth18	mental_heth18	mental_heth16	mental_heth16
	Estimated results of the first stage			
exercise	-0.070***		-0.020*	
	(0.013)		(0.012)	
d_alone		-0.527***		-0.370***
		(0.044)		(0.043)
Control variable	Yes	Yes	Yes	Yes
Province fixed effect	Yes	Yes	Yes	Yes
Cragg-F	56.73	60.25	42.53	45.05
Observations	9291	9291	9291	9291
R2	0.19	0.20	0.15	0.16
	Estimated results of the second stage			
	(1)	(2)	(3)	(4)
variable	family_income	family_income	family_income	family_income
mental_heth18	-0.095**	-0.077***		
	(0.047)	(0.018)		
mental_heth16			-0.334*	-0.110***
			(0.252)	(0.028)
Control variable	Yes	Yes	Yes	Yes
Province fixed effect	Yes	Yes	Yes	Yes
Observations	9291	9291	9291	9291
R2	0.36	0.39	0.23	0.32

Note: The values in parentheses are standard errors, and *, **, *** indicate significance levels of 10%, 5%, and 1%, respectively.

4.3 Robustness test

The above analysis preliminarily confirms that the degree of mental health has a positive impact on family income. In order to verify its robustness, this article uses the following two methods. First, referring to the practice of Bogan and Fertig [9], this article sets a dummy variable of family mental health as the core explanatory variable, denoted as un_health. If the person in charge of family finance answers the 8 mental health questions asked, 5 questions or more scored 2 to 3 points, then the dummy variable value is 1, otherwise it is 0. The model (1) and model (2) in Table 5 give the estimation results of the mental unhealthy dummy variable on the family income. Secondly, in the CFPS interview, the interviewer will score some subjective questions in the questionnaire based on their own feelings to ensure the quality of the data. This article deletes the interviewer's "respondent's ability to understand", "respondent's level of cooperation with the survey", "respondent's intelligence level", "respondent's interest in the survey", and "respondent's response to the survey". For samples whose scores are lower than the median (2 points), the remaining samples are re-analyzed, and the model in Table 5 (3) and model (4) give the result of re-estimating the degree of mental health on family income.

Table 5. Robustness test results.

	(1) family income	(2) family income	(3) family income	(4) family income
un_health18	-0.230***			
	(-3.99)			
un_health16		-0.176***		
		(-2.62)		
mental_heth18			-0.017***	
			(-4.88)	
mental_heth16				-0.019***
				(-4.78)
Control variable	join	join	join	join
Province difference	control	control	control	control
Sample size	9291	9291	4436	4436
R^2	0.417	0.416	0.436	0.437
F	184.097	184.317	103.805	103.755

Note: The brackets are the robust standard errors of the estimated coefficients; *, **, *** represent that the coefficients are significant at the levels of 10%, 5%, and 1%, respectively.

Compared with the estimated results of the model (1)(2) in Table 5, the estimated results of the model (1)(2) in Table 5 have a slight change in the value of the division coefficient, and their significance and signs are highly consistent with the estimated results of the model (1). Although the coefficient has increased significantly, it may be due to the fact that under the pressure of today's society, mentally unhealthy families account for the majority, and the change in data has increased the impact on family income. It can be seen that the statistical results of different regressions are all significant at the 10% level. Compared with the estimation results of the full sample model (1) and model (2) used in Table 2, the estimated coefficients of the model (3) and the model (4) that use more reliable samples are almost the same. The above analysis results show that the positive effect of mental health on family income is stable.

4.4 The mediating role of cognitive ability

Cognitive difference effect means that different mental health status affects the individual's optimization ability by changing the individual's cognitive ability, and then affects the individual's financial management and labor ability. Cognitive ability refers to the ability of the human brain to process, store and extract information. It describes the "intrinsic" ability of a person in terms of language ability, computing ability and memory. Existing studies have shown that there is a significant positive relationship between cognitive ability and household income, which has a certain degree of influence on individuals' participation in financial management and processing of intensive information [26].

In order to test the mediating role of cognitive ability, the article first examines the relationship between cognitive ability and family income. If cognitive ability has a significant impact on family income, it will further examine the impact of mental health on cognitive ability, that is, verify that the greater the degree of mental health Does the poorer respondent have worse cognitive abilities? If both are significant, mental health has an indirect impact on family income through the mediating effect of cognitive ability [27]. It is worth pointing out that, limited to the method of constructing cognitive ability variables, this article can only test the mediating role of cognitive ability in the influence of mental health on family income. The basic process of the mediation effect test is as follows:

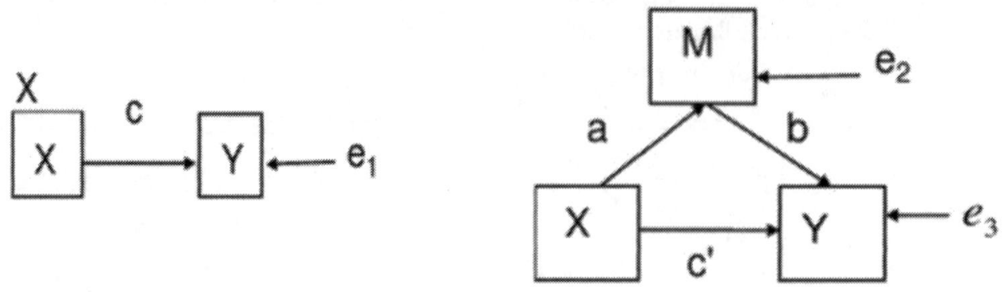

Fig.1. Schematic diagram of the mediation effect test.

In Figure 1, the explanatory variable X influences the explanatory variable Y through the intermediate variable M. In the analysis of this article, X is the family's mental health, M is the cognitive ability, and Y is the family income. In Figure 1, c is the total effect of mental health on family income, which has been estimated by Equation (2) above. In Figure 1, a is the influence of mental health on cognitive ability. When both X and M are continuous variables, they can be estimated by the following equation (6):

$$M = aX + \varepsilon_1 \tag{6}$$

If X, M, and Y are all continuous variables, and can be estimated by the following equation (7):

$$Y = c'X + bM + \varepsilon_2 \tag{7}$$

The model in Table 6 is the estimated result of the above equation. Due to space limitations, the marginal effect results of other control variables are omitted in this article.

Table 6. Regression results of the mediating effect of cognitive ability.

	x2cog equation （6）		family_income equation （7）	
	Marginal effect	Robust standard error	Marginal effect	Robust standard error
mental_heth18	-0.062***	0.004	-0.016***	0.002
cog			0.051***	0.006
Control variable	join		join	
Province fixed effect	Control		Control	
Cragg-F	361.28		185.95	
Observations	9291		9291	

Note: ***, **, *represent that the coefficient is significant at the level of 1%, 5%, and 10%

It can be seen from Table 6 that the estimated coefficient a of mental health on family income and the coefficient b of influence of cognitive ability on family income are both significant at the 1% level. After controlling for cognitive factors, the direct impact of mental health on family income c is significant at the 1% level, and the direct effect of mental health on family income is -1.5%,use[(c-c')/c]Indirect calculation is available, The mediating effect of risk attitudes accounted for approximately 16.96% of the total effect. The mediating effects of cognitive ability have passed the Sobel test and Bootstrap test respectively, as shown in Table 7. This result means that the mediating role of cognitive ability in the influence of mental health on family income is established, which verifies Hypothesis 2 of this article.

Table 7. The mediation effect test of cognitive ability.

	Coef	StdErr	Z	P>|Z|
Sobel	-0.003	0	-7.493	0

Bootstrap	Observed Coef.	Bias	Bootstrap Std.Err.	[95%Conf.Interval]
ind_eff	-0.003	0	0	[-0.004 -0.002]
dir_eff	-0.0156	0	0.003	[-0.021 -0.010]

4.5 Analysis of heterogeneity of household income level

In the "Healthy China 2030" plan promulgated and implemented in 2016 by the Central Committee of the Communist Party of China and the State Council, it is clearly stated that "We must adhere to the joint contribution and sharing, and the health of the whole people... The health problems of low-income groups and other key groups". Generally speaking, personal background often determines the degree to which an individual is affected by mental health, and understanding the sensitivity of different groups' family income to mental health is conducive to more accurate formulation of relevant policies to carry out social security work for specific groups of people. Therefore, in order to further analyze the impact of mental health factors on the household income of various income classes in our country and conduct a more detailed study of the different levels of data, the following quantile regression method is used to establish a model, and the modeling process is matched with the income class division 3 etc. Select quantile points as 0.25, 0.5, 0.75. See Table 8 for details.

Table 8. Quantile regression results.

	q25	q50	Q75
mental_heth18	-0.029***	-0.015***	-0.011***
	(-10.02)	(-6.16)	(-4.05)
gender	-0.040*	-0.036**	-0.020
	(-1.93)	(-2.12)	(-0.94)
age	0.051***	0.036***	0.031***
	(8.66)	(7.95)	(8.64)
age2	-0.001***	-0.000***	-0.000***
	(-10.54)	(-8.61)	(-8.30)
edu	0.132***	0.123***	0.115***
	(12.86)	(15.21)	(12.41)
family_size	0.155***	0.151***	0.130***
	(22.80)	(27.21)	(25.66)
registered_res	-0.369***	-0.315***	-0.218***
	(-10.26)	(-12.34)	(-9.63)
heth	-0.021**	-0.024**	-0.029***
	(-2.01)	(-2.45)	(-2.91)
house_value	0.231***	0.217***	0.213***
	(19.85)	(23.74)	(36.42)
Inter_ac_inf	0.038***	0.047***	0.048***
	(4.90)	(6.47)	(6.54)
social_contact	0.007	0.010**	0.003
	(1.04)	(2.10)	(0.51)
social_trust	0.003	0.003	0.010**
	(0.60)	(0.74)	(1.97)
Constant term	6.449***	6.910***	7.506***
	(22.97)	(41.56)	(36.03)
Sample size	9291	9291	9291
R^2	0.31	0.28	0.26

Note: The values in parentheses are t-values; ***, **, * indicate statistical results are significant at 1%, 5%, and 10% respectively

use“test[q25=q50=q75]:mental_heth18”To test whether the coefficients of mental health (mental_heth18) are the same in several quantile regressions, the following results are obtained:

(1)[q25]mental_heth18-[q50]mental_heth18=0
(2)[q25]mental_heth18-[q75]mental_heth18=0
F(2,9254)=16.38
Prob>F=0.0000

The result shows that the regression coefficients of the above 3 quantiles are significantly different, which is consistent with hypothesis 3. It can be seen from Table 8 that the mental health elasticity coefficient of different quantile groups fluctuates greatly, indicating that the degree of mental health impact on family income is relatively different between groups. Basically, as the family income class decreases, the greater the impact of mental health status, it can be considered that the lower-income families are more likely to enter the vicious circle of mental health and low income. High-income families may use medical treatment, entertainment and other methods to relieve and improve their mental health.

5 Conclusion

Based on the data from the China Family Tracking Survey (CFPS) in 2018, this article empirically studies the relationship between mental health and family income, and further explores the mediating role of cognitive ability. The empirical results show that the level of mental health is significantly positively correlated with the level of family income. For every 1% decrease in the level of family mental health as measured by the CES-D8 score, the family income level will increase by 1.7%-1.9%, compared with Chushchyan and others based on Medical- Expendi-ture Panel Survey data explored the impact of physical and mental health on the annual salary of American residents. For every 1% increase in mental health, the annual salary increased by 0.13% (0.09%~0.16%) [28]. Although the whole family is studied here, the impact of mental health problems in our country seems to be more serious. After redefining the mental health indicators, deleting unreliable samples, adding the pre-determined explanatory variable 2016 mental state level, etc., using the filtered loneliness as an instrumental variable, the two-stage least squares also got the same conclusion. The above results are still robust. In addition, the change of cognitive ability is an important channel through which mental health affects family income. Mental-healthy families have better cognitive and judgment skills, which can not only obtain more human capital and wage income in the labor market, but also may increase financial income. The result of indirect calculation shows that the mediating effect of cognitive ability is about 16.96% of the total effect of mental health affecting family income level. In terms of heterogeneity analysis, it is studied that the impact of mental health problems at different family income levels is different. The marginal effect of the mental health impact of low-income families is as high as -2.9%. The main reason for the above differences is that Due to differences in income levels, different families face different pressures, environments and endurance, social needs, and the cost of alleviating and controlling mental illnesses, which ultimately leads to the heterogeneity of the impact of mental health problems on family income.

In the process of rapid economic growth in China, mental health status has a significant impact on family work income. Under great psychological pressure in work and life, more and more residents often suffer from anxiety and depression. Therefore, in order to increase labor productivity, improve financial management capabilities and promote family income, the mental health of national families urgently needs to be paid attention to. This article believes that the government needs to properly address the negative impact of mental health problems on individuals, families and society, and actively promote the popularization of the mental health of the whole people, especially the need to focus on low- and middle-income family groups, so as to prevent mental illness in the new era of building a well-off society in an all-round way. The resulting family poverty problem has effectively improved the happiness of the people. In addition, it is recommended that my country establish a sound family mental health counseling system, so that families can receive psychological help in a timely and effective manner, improve their mental health, and promote family income.

References

1. Mushkin S. Health as an Investment[J]. Journal of Political Economy, 1962, 70(5): 129-157.
2. World Health Organization. World Health Statistics 2019: Monitoring health for the SDGs[R]. Geneva: WHO, 2019.
3. Vigo D, Thornicroft G, Atun R. Estimating the true global burden of mental illness[J]. The Lancet Psychiatry, 2016, 3(2): 171-178.
4. China National Mental Health Development Report (2019～2020) [R]. Institute of Psychology, Chinese Academy of Sciences and Social Science Literature Publishing House, 2021.Frijters P, Johnston D, Shields M. Mental Health and Labor Market Participation: Evidence from IV Panel Data Models[J]. IZA Discussion Papers, 2010, 4883.
5. Nikki Bond and Conor D'Arcy. Mind the income gap[R]. Money and mental health policy institute, 2020.
6. Thomas, Patricia A., Hui Liu, and Debra Umberson. Family relationships and well-being[J]. Innovation in aging, 2017, 1(3): 1-11.
7. Larson, Reed W., and David M. Almeida. Emotional Transmission in the Daily Lives of Families: A New Paradigm for Studying Family Process[J]. Journal of Marriage and Family, 1999, 61(1): 5–20.
8. Vicki L. Bogan, Angela R. Fertig, Portfolio Choice and Mental Health[J]. Review of Finance, 2013, 17(3): 955–992.
9. Richardson T, Elliott P, Roberts R, et al. A longitudinal study of financial difficulties and mental health in a national sample of British undergraduate students[J]. Community mental health journal, 2017, 53(3): 344-352.
10. Ozawa S, Laing S K, Higgins C R, et al. Educational and economic returns to cognitive ability in low-and middle-income countries: A systematic review[J]. World Development, 2022, 149: 105668.
11. Dajun Lin, Randall Lutter, Christopher J. Ruhm, Cognitive performance and labour market outcomes[J]. Labour Economics, 2018, 121-135.
12. Yu Yijia, Cognitive Ability and Family Asset Selection[J]. Economic Research, 2014,12: 132-142.
13. Tyng C M, Amin H U, Saad M N M, et al. The influences of emotion on learning and memory[J]. Frontiers in psychology, 2017, 8: 1454.
14. Yu Guoliang, Dong Yan. Current status, hotspots and development trends of mental health research in my country[J]. Educational Research, 2014, 6: 97-102.
15. Turvey C., R. Wallace and R. Herzog. A Revised CES－D Measure of Depressive Symptoms and a DSM-Based Measure of Major Depressive Episodes in the Elderly[J],

International Psychogeriatrics,1999, 11(2): 139-148.

16. Li Tao, Zhu Junbing, Fu Lin. Are smart people more willing to start a business? ——Empirical findings from China[J]. Economic Research, 2017, 594(03):91-105.
17. Zhou Yang, Wang Weihao, Liu Xuejin. Cognitive Ability and Financial Exclusion of Chinese Households——An Empirical Study Based on CFPS Data[J]. Economic Science, 2018, 223(01): 96-112.
18. Zhang Chuanchuan. An Empirical Analysis of the Impact of Health Changes on Labor Supply and Income[J]. Economic Review, 2011(04): 79-88.
19. Wang Xiuzhi, Yi Ting. The income effect of healthy human capital[J]. Journal of Capital University of Economics and Business, 2017, 19(04): 20-26.
20. Xu Qiuyan, Zhang Qiuwei. Family education level, health level and income level: An empirical analysis based on the follow-up survey of Chinese families[J]. Research World, 2018, 299(08): 19-28.
21. Huang Yun, Ren Guoqiang, Zhou Yunbo. The impact of income inequality on the physical and mental health of rural residents: an empirical analysis based on CGSS2015 data[J]. Agricultural Technology and Economy, 2019(03): 25-37.Hamilton V, Merrigan P, Dufresne E. Down and out: estimating the relationship between mental health and unemployment[J]. Health Economics, 1997, 6(4): 397-406.
22. Chatterji P, Margarita a, Lu M, et al. Psychiatric Disorders and Labor Market Outcomes: Evidence from the National Latino and Asian American Study[J]. Health Economics, 2007,16(10): 1069-1690.
23. Stock J, Yogo M, Wright J. A Survey of Weak Instruments and Weak Identification in Generalized Method of Moments[J]. Journal of Business and Economic Statistics. 2002, 20: 518-529.
24. Gregg, Paul and Lindsey Macmillan. Family income, education and cognitive ability in the next generation: exploring income gradients in education and test scores for current cohorts of youth[J]. Longitudinal and life course studies, 2010, 1: 259-280.
25. Wen Zhonglin, Fang Jie, Shen Jiaqi et al. Review of domestic research on psychological statistics methods in the new century[J]. Advances in Psychological Science, 2021, 252(08): 1331-1344.
26. Chushchyan V H, Campbell JD, Libby A M. Physical and mental health as important inputs into wage function[J]. Value in Health, 2011, 14(3): A193.

Fashion Design and Aesthetic Embodiment Based on Virtual Reality Technology

Jiayi Shi

School of Gemology, Shaanxi Institute of International trade & commerce, Shaanxi, Xi'an, 710000

Abstract. With the continuous development and improvement of virtual reality technology, the technology has been widely used to improve the overall level of fashion design, vivid aesthetics to create a good foundation. By giving full play to the advantages of virtual reality technology application, innovating fashion design scheme, optimizing fashion design concept, annotating aesthetic space, enriching aesthetic embodiment, helping fashion design to realize overall innovation. Based on the application of virtual reality technology, this paper puts forward some reasonable suggestions on fashion design and aesthetic embodiment.
Keywords: Virtual reality technology; fashion design; aesthetic embodiment; design innovation

With the development of science and Technology, computer technology has been able to construct virtual reality scenes, and people can experience the sense of being in the scene from visual, auditory and other sensory aspects. The development of virtual reality (VR) technology is greatly promoting the diversity of fashion expression, breaking the traditional limitation, redefining the relationship between the physical clothing and the virtual digital world, and realizing the digitization of the physical.

1 Fashion design theory based on virtual reality technology

Virtual reality technology is based on the premise of computer-supported simulation technology. It forms a virtual environment, virtual process and virtual product through Unified Modeling of design, processing, assembly, maintenance, etc. , it consists of hearing, sight, touch, smell and interaction. Based on virtual reality technology, mechanical products as the object of the design means. Virtual reality is not only a man-machine interactive system based on natural way, but also a virtual environment generated by computer, and through a variety of sensing devices, the user has the feeling of being in the scene.

Virtual design is the virtual realization of the whole process of product from concept design to put into use in the virtual environment constructed on the computer. Its goal is not only to simulate and visualize the material form and manufacturing process of the product, and it is to predict, evaluate and optimize the performance, behavior and function of the product and the implementation plan in each stage of the product realization.

The uniqueness of virtual reality technology lies in that it bridges the distance between reality and art, transcends technology itself and has its own aesthetic value. Starting from the media digital ontology of virtual reality, this paper expounds its non-material aesthetic characteristics, and starting from the three major characteristics of virtual reality technology: conceiving, immersing and interacting, the characteristics of hyper-real aesthetics, immersion aesthetics and non-linear aesthetics of virtual reality art are summarized. With the coming of digital age, the boundary between virtual network and reality is becoming more and more blurred. Fashion is the epitome of the evolution of the times. Designers use modern technology to fully digitize fashion, creating contactless virtual fashion designed for the Internet age.

2 Fashion design aesthetic system based on virtual reality technology

Artistic Beauty is one of the forms of beauty, which comes from the aesthetic cognition of the artist's subject and is the reflection of realistic beauty in art. The beauty of art originates from the beauty of reality, but at the same time it is higher than the beauty of reality. The beauty of art includes the beauty of artistic content and the beauty of artistic expression. The composition of artistic content contains not only the image of the universe, but also the cultural aesthetic factors. The artistic beauty in the virtual reality system is also created by the designer in the virtual reality system after realizing the true meaning of art, in order to meet the users' Aesthetic Vision, the designer is creating based on real-world data. The so-called multi-perception refers to the perception of hearing, touch, motion, taste, smell, etc. . The beauty of art in the virtual reality system can arouse people's sense organs, it makes you feel like you're in a virtual reality system. Virtual reality art has interactive experience. Virtual reality technology itself is a human-computer interaction technology, and its interactivity requires participants to operate the virtual objects in the virtual environment. Human, as the subject of virtual environment, is the driving force for the continuous improvement of the interaction of virtual reality system. The interaction of virtual reality technology emphasizes the real-time interaction between human and computer. Let people with the help of virtual reality technology from the daily real life, into a more free virtual environment. The artistic transcendence of the virtual reality system is also reflected in its ability to break through the constraints of space-time and material conditions, making people feel what they can not experience in real life.

The beauty of technology is an independent aesthetic applied discipline, which is the product of highly developed modern productive forces. The use of technical beauty can promote the aesthetic creation of products, which is the aesthetic value of technical activities and works of art. The beauty of technology in virtual reality needs the support of computer hardware and software, and should pay attention to the technology of environment modeling, real-time rendering and collision detection. The technology of environment modeling includes geometry modeling, physics modeling and action modeling. Geometric modeling is the foundation of Environment Modeling in virtual environment, which mainly deals with and represents the geometric information of object's shape and mechanism. In the process of geometric modeling, special attention should be paid to the spatial proportion and shape of objects, in order to lay the foundation for the next stage of physical modeling. In addition, the details of things should be properly highlighted in the geometric modeling. Physical modeling is just modeling the material and color of objects in the virtual environment. If we want to reflect the real things in the virtual reality system, we must attach great importance to the material and color of objects, the color of the object is the key factor that affects people's visual feeling, directly affects people's aesthetic psychology and affects people's emotion. Therefore, in the creation of virtual environment, we should pay special attention to the unified effect of color.

Design Technology provides new design thinking for modernization. In the field of traditional art design, design technology is often the most difficult problem to break through. Many times, the inspiration of good art design is not enough to support the creation of the technical and had to complete expression, this kind of helplessness is the most difficult for every artist to accept. Traditional art design is often hand-written and drawn, and it is easy to create for small-scale design, but for super-large-scale art design, it requires a lot of manpower and material resources, and in the difficult and complicated creative process, any calculation, handwriting, or description of the slightest problem is likely to bring irreparable damage to the art and design. In the new media era, various new technologies and technologies emerge one after another, such as digital technology, Big Data Technology, virtual reality technology, etc.. The invention and innovation of these technologies provide solid technical support for art design, at the same time, artistic creation has broken the traditional single-field design of the pure paper media. The modernization of design technology has promised the grand vision of "nothing is impossible, only unexpected" for the artist, which makes the artist's thinking burst out with rich creative energy and artistic inspiration.

3 Fashion design based on virtual reality technology

Graphic Design, as the most commonly used design method in the new media era, has a high demand on color design. In the new media era, choosing the right, harmonious and distinct color design is the direct means to deepen the audience's understanding of information, which is the basic aesthetic requirement for the creators in the current new media era. The correct and proper use of color matching design aids the entire creative design to enhance the design style, highlight the design theme, and effectively set off the design atmosphere. This type of color design can quickly attract the audience's attention, enhance the attention of the work, at the same time can make users in the influence of color to achieve a more background color with the color design. For example: the original desktop of Microsoft system computer, adopted the color collocation effect of green field and blue sky, this kind of color collocation design makes the feeling that the user opens the computer is tranquil and peaceful, can play the role of soothing mood.

Perfect structural design is the focus of object or space design. For the structural design in the new media age, on the basis of aesthetic creation and practical use, the artistic creator must consider the emotional needs of the audience and optimize the structure of the designed works, ensure that the user can use the work in the simplest, most reasonable and most appropriate way. The media plays a decisive role in the development of human information and the way of thinking, and in the practice of human society, it has produced relevant standards of behavior and ideas, for us to win a technology and humanities complement each other in a larger space-time. The holographic text form, mimicry communication, interactive body narrative and three-dimensional aesthetic transformation of virtual reality art have once again aroused the new thinking of the "Meta-problem" of contemporary art.

Information symbol is the most concise form of information transmission and language expression. The shape and form of the symbol reflect the designer's intention directly. It should be noted that each audience's environment, education, religious beliefs and other elements are different, resulting in different people's understanding of the same symbols and analysis of different answers. Therefore, in the design of information symbols in the new media age, the creator of art must take full account of the audience's contact environment, education level, religious belief and other factors, actively avoid symbolic information that may mislead the audience when designing. Due to the complexity of information in the new media age, many viewers can not devote too much energy and spirit to a single symbol. In order to improve the audience and effect of symbol design, symbol design should be simple and clear, with clear implications, the design must not be too complex and difficult to understand, otherwise the audience will soon lose interest in watching and understanding, which puts forward higher requirements for designers in the new media age. Designers must draw from the information itself to create a simple and clear symbolic language, and can accurately and clearly express the content and meaning of the information itself, at the same time, we should pay attention to the structure and aesthetics of the design of the symbol itself.

Virtual reality technology has created a powerful "hyper-real" space, in the path of complex link technology, the means of expressing ideas in artistic texts are gradually changing from planar images to multimedia images, non-media images and other media forms, and they embody the distinctive characteristics of post-modern culture, such as immersion, collage and disorder, there is no depth, no center, no meaning. Therefore, some scholars say: "the advent of the image society has reconstructed the Cultural Order of human society: People's Way of life and the way in which people receive information, enter the imaginary world and the spiritual world are gradually filled with and surrounded by images." Indeed, we are in an age when images dominate the media, and the expansion of cultural consumption of digital images has led to the reconstruction of cultural order, the aesthetic turn from "literary figure" to "Figure Shadow" in the visual context can not be ignored.

Virtual reality allows cross-media and cross-cultural integration of digital art and other visual cultures, blurring the line between art and non-art, it also makes the appearance of virtual reality art once again put the "Meta-problem" of the essence of art in front of the audience, "what is art", "how art is created", "what is art?" And so on, it may rekindle interest in the art of virtual reality. In a sense, all arts are languages, but they are only languages in a certain sense. They are not the language of literary symbols, it is the language of perceptual symbols." Images have an intuitive ideographic function, and also belong to a form of language symbols. In contrast, the visual presentation of the plane image in the spiritual dimension and poetic feeling is not as strong as the aesthetic experience of the three-dimensional visualization of the virtual image. Virtual reality art text combines picture, text, sound, image and other media, and presents the external image and internal consciousness of art in harmony. Multi-media fusion has once again deepened the depth and breadth of text aesthetic value.

The rapid development of virtual reality technology has promoted the essence transformation and value derivation of visual art, challenged the traditional aesthetic system structure of art, and also affected people's overall understanding of art, at the same time, it creates the new possibility of "digital existence" and changes the living space and thinking dimension of human beings. As a connecting way of "body-like world" and a new force of constructing and deconstructing social cultural phenomena, virtual reality has become the representative of visual aesthetic culture, supported by a series of new media technologies, the exhibition of virtual reality art aesthetics is more diversified than previous digital images, and expands the boundary of digital visual aesthetic culture. Compared with the one-way graphic film and television art, virtual reality art is more dynamic, attracting more netizens to participate in it, stimulating the enthusiasm of the general public for virtual reality art. The "Internet +" series of image art has completed the challenge to the traditional art, pushing the art creation from the elite to the public, while the virtual reality art has once again challenged the traditional mechanism of art, further deepened the new media art system cognition.

4 Conclusion

If virtual reality art is to maintain the artistic individuality of its aesthetic creation, it is indeed necessary to integrate the positive factors that permeate the traditional artistic discourse system and artistic aesthetic concepts to a certain extent, at the same time, it emphasizes the quality of its own artistic creation and tries to combine and penetrate the quality of the two artistic systems. As far as virtual reality art is concerned, its imitative "hyper-real" expression and flexible aesthetic participation are more suitable for the public's consumption habits of visual culture, with the help of its tension-rich and flexible artistic elements, to constantly improve their aesthetic quality, so that different art forms in the integration of infiltration to promote the contemporary art sector as a whole sound development.

References

1. Niu Hongying. A contemporary approach to traditional aesthetics under the condition of virtual reality technology [J]. Chinese Literature and Art Review, 2017, (12): 13-21.
2. Gong Ting. On the Evolution of Aesthetic Subject and Aesthetic Object in Virtual Society [J]. Guizhou Social Sciences, 2017, (12): 93-98.
3. Liu Zhejun, Yao Yao. On the Aesthetic Art Characteristics and Value of Virtual Reality Animation [J]. Journal of Harbin University, 2017, (10): 122-125.
4. Chen Qi. Research on experiential display design based on virtual reality technology [J]. Art Education, 2017, (13): 201-202.
5. Jiang Shen. Immersion: Aesthetic Origin, Innovation and Technical Approach of Virtual Reality [J]. Art Science and Technology, 2018, (12): 9-10.

AUTHOR INDEX

EDP Sciences – Web of Conferences
17, Avenue du Hoggar
Parc d'Activité de Courtabouf
BP 112
F-91944 Les Ulis Cedex A
France

ISBN 978-1-7138-5362-6

CURRAN ASSOCIATES INC.
proceedings.com